Vegetable Seed Processing

Vegetable Seed Processing

Rakesh C. Mathad *(Assistant Professor)*
Seed Science and Technology at Seed Unit
University of Agricultural Sciences, Raichur, Karnataka

Basave Gowda *(Professor)*
Seed Science and Technology and Special Officer (Seeds)
University of Agricultural Sciences, Raichur, Karnataka

S. B. Patil *(Assistant Professor)*
Seed Science and Technology, College of Agriculture, Bheemaryanagudi
University of Agricultural Sciences, Raichur, Karnataka

NEW INDIA PUBLISHING AGENCY
New Delhi – 110 034

NEW INDIA PUBLISHING AGENCY

101, Vikas Surya Plaza, CU Block, LSC Market
Pitam Pura, New Delhi 110 034, India
Phone: + 91 (11)27 34 17 17 Fax: + 91(11) 27 34 16 16
Email: info@nipabooks.com
Web: www.nipabooks.com
Feedback at feedbacks@nipabooks.com

© 2015 : Rakesh C. Mathad

ISBN : 978-93-85516-03-0

All rights reserved, no part of this publication may be reproduced, stored in a retrieval system or transmitted in any form or by any means, electronic, mechanical, photocopying, recording or otherwise without the prior written permission of the publisher or the copyright holder.

This book contains information obtained from authentic and highly regarded sources. Reasonable efforts have been made to publish reliable data and information, but the author/s, editor/s and publisher cannot assume responsibility for the validity of all materials or the consequences of their use. The author/s, editor/s and publisher have attempted to trace and acknowledge the copyright holders of all material reproduced in this publication and apologize to copyright holders if permission and acknowledgements to publish in this form have not been taken. If any copyright material has not been acknowledged please write and let us know so we may rectify it, in subsequent reprints.

Trademark notice: Presentations, logos (the way they are written/presented) in this book are under the trademarks of the publisher and hence, if copied/resembled the copier will be prosecuted under the law.

Composed, Designed and Printed in New Delhi, India

Dr. P. M. Salimath
Vice Chancellor

University of Agricultural Sciences,
Lingasugur Road, Raichur – 584 104.
Off   : 08532 221 444
Mob.   : 94806 96300
Fax   : 08532 220 444
E-mail : vcuasraichur10@rediffmail.com
Web.   : uasraichur.edu.in

FOREWORD

India is the second largest producer of vegetables after China and also has a unique distinction of one of the most preferred destinations for vegetable seed production. The south India becomes hub for vegetable seed production of tropical vegetables in south-east Asia. This is because of presence of ideal weather conditions, soil and skilled labour force. The cost of production per kilogram is very low compared to Europe and Americas. Some of the well known seed corporations of the world are operating in this region with annual grower payments exceeding 100 million USD. This activity created jobs and has positive socio-economic impact.

The seed processing is an important activity after production and to be taken care with utmost importance at the shop floor. Since there is wide diversity of vegetable hybrids or varieties, their processing requirements also vary. Each hybrid or variety may behave differently with respect to screens, processing machines, seed treatment etc. In this background an effort to give detailed information on vegetable seed processing, this book can be a valuable reference to students, academic and industry personnel associated with processing.

I appreciate an effort of Mr. Rakesh C. Mathad, Dr. Basave Gowda and Dr. S B Patil, Seed Unit, UAS, Raichur in compiling such an valuable resource book for the shop floor.

I hope this publication would be useful to the various personnel involved in vegetable seed production, processing, testing and certification.

April, 2015

(P.M. SALIMATH)
Vice-Chancellor
University of Agricultural Sciences,
Raichur-584 102.

Contents

Foreword v

1. **Introduction** 1
   1.1 Introducution 1
   1.2 Seed processing of tropical vegetables 3
   1.3 Flow chart of seed processing 4

2. **Harvesting and Seed Pre-conditioning at Farmers Level: Two Case Studies of Tomato and Sponge Gourd** 5
   2.1 Harvesting and Pre-conditioning 5
   2.2 Harvesting and Pre-conditioning in tomato (Wet Seed Type) ... 6
   2.3 Harvesting and Pre-conditioning in sponge gourd (Dry Seed Type) 7

3. **Processing Line and Equipments in Vegetable Seed Processing** 9
   3.1 Principle of seed processing 9
   3.2 Requirement in seed processing 10
   3.3 Types of materials removed during seed processing 10
   3.4 Sequence of operation in seed processing 10
   3.5 The flow chart illustrating the types of materials removed from harvested produce during processing. 11
   3.6 Seed processing equipments 12
   3.7 Seed packaging equipments 27

4. **Vegetable Seed Processing** 31
   4.1 Beet Root 31
   4.2 Brassicas (Cabbage, Cauliflower, Knol Khol) 32
   4.3 Carrot 33
   4.4 Caraway 34

4.5	Celery	35
4.6	Chicory	36
4.7	Coriander	37
4.8	Sweet Corn	38
4.9	Gherkins or Pickling Cucumber	38
4.10	Cucumber (General)	39
4.11	Pumpkin	40
4.12	Cumin	41
4.13	Dill	42
4.14	Eggplant or Brinjal	43
4.15	Fennel	44
4.16	Fenugreek	45
4.17	Endive	45
4.18	Leek	46
4.19	Lettuce	47
4.20	Watermelon	48
4.21	Okra	49
4.22	Onion	50
4.23	Parsley	50
4.24	Pepper (Hot Pepper or Chillies and Capsicum)	51
4.25	Radish	52
4.26	Rhubarb	52
4.27	Rosemary	53
4.28	Rue	53
4.29	Spinach	54
4.30	Squash or Zucchini	55
4.31	Tomato	56
4.32	Turnip	56

5. Seed Processing for Improving Quality and Purity ... 59

5.1	Processing to improve seed genetic purity	59
5.2	Processing to improve seed germination or vigour	63
5.3	Liquid density separation	63

Contents / ix

6. **Seed Blending** 67
    6.1   Why seed blending 67
    6.2   Methods in seed blending 68
    6.3   Principles in seed blending 70

7. **Seed Treatment** 73
    7.1   Benefits of seed treatment 73
    7.2   Types of seed treatment 74
    7.3   Scope for seed treatment- Must Do's 74
    7.4   Compounds used for seed treatment 75
    7.5   The following are the different combinations of polymer and chemicals for seed treatment. 77
    7.6   Seed treatment methods 78
    7.7   Equipment for seed treatment 78
    7.8   Wet seed treatment 80
    7.9   Seed quality enhancement techniques 80
    7.10 Precautions in seed treatment 84
    7.11 Causes of poor treatments 84
    7.12 Cost of seed treatment 85

8. **Seed Drying** 87
    8.1   Principles of drying 88
    8.2   Methods of seed drying 92

9. **Seed Storage** 97
    9.1   Factors affecting shelf life or longevity seeds 98
    9.2   General principles of seed storage 103
    9.3   Types of vegetable seed storage 104
    9.4   Factors to be considered in storage 108

10 **Seed Inventory Management Systems** 111
    10.1 SAP Basic information & structure 111
    10.2 SAP structure 111
    10.3 SAP functional modules 112
    10.4 Screen structural elements 113

11. Guidelines for Quality Vegetable Seed Production 119
    11.1 Planning 119
    11.2 Stock seed handling 119
    11.3 Greenhouse/Insect proof net production 120
    11.4 Direct sowing 123
    11.5 Transplanting in Net or open field 124
    11.6 Roguing (Field inspection) 124
    11.7 Maturation or Harvesting 128
    11.8 Seed receipt at Processing plant 129

12. Glossary 135

CHAPTER - 1

Introduction

1.1 Introducution

Growing of vegetables is one of the most profitable allied activities of agriculture especially in small-medium land holdings and with limited water resources. Use of improved varieties and hybrids is very critical to make smaller land holdings profitable. Seeds of improved varieties and hybrids not only increase income to the tune of more than 30% but also reduce the use of plant protection chemicals. India being the second largest producer of vegetables in the world with 6.2 m ha of land and 85m MT production-use of improved seed is very important. The annual requirement of vegetables is estimated to be over 135m MT which can be only achieved through use of quality seed. India is predominantly vegetarian country and depends upon vegetables for their nutrients and minerals.

Vegetable cultivation is labour intensive, gives high returns and ensures nutritional security. To ensure these the supply of quality seed is very vital. Apart from seed production, its processing for seed quality to give the end user is of high importance. Unlike seed processing of field crops, vegetable seed processing is very technology driven since the volume is less and involve costly material. Improved processing efficiency brings more income for the grower and seller. The vegetable seed being low volume needs better technology for upgrading, size grading, blending, conditioning, processing, treatment and packaging. Vegetable seed processing is set of activities as mentioned here each of having a special focus on the quality.

Here are some of the terms in vegetable seed processing which are used for specific functions:

1.1.1 Seed Cleaning: involves cleaning of debris, infected or infested material immediately after harvest.

1.1.2 Pre-conditioning: any activity intended towards making seed ready for processing like removal of undersized seeds, inert matter and treatments like acid treatment or wet seed treatment. This activity is done at seed grower level.

1.1.3 Conditioning: means removal of chaff, immature seeds, weed seeds, inert matter and other crop seeds at the shop floor. Seed conditioning covers operations like drying to optimum moisture levels, treatment, packaging or labelling. Here use segregation of seeds based on size, shape, colour or density is very important since different crops have different conditioning requirements.

1.1.4 Upgrading or Size Grading: processing of seed lot to obtain uniform quality in terms of size, shape or texture or any operation which will change the purity or germination of the seed and require retesting to determine the quality of the seed. Also improvement of a seed health by various heat or hot water treatment also done here. Improvement of quality by using methods like liquid density is also practiced here. Seed blending to improve quality parameters like germination is also done following some rules.

1.1.5 Processing: processing of seed lot refers to conditioning plus sampling, blending, treatment, packaging and labelling also. Sometimes processing covers germination before packing and guard sampling. Vegetable seed processing is used as a collective term combining all the above activities in a shop floor of a processing assembly. Specific equipments were used for different seed lots and each lot behaves distinctly when it comes to physical characteristics.

There are different processing methods for different class of seed based on the standards or specs. The processing requirement of commercial seed is different from breeder stock seed and also depending upon the target market if the seed is being exported. The classification of seed classes based on the Seed Act. 1966 & Rules, 1968 and processing requirement as follows:

Breeder Seed and Stock Seed: Processing for this class of seed is highly specialized and seed need to be 100 % pure in terms of germination, genetic purity and seed health-free from any seed borne diseases.

Foundation Seed: Here a flexibility of <1% in terms of quality parameters as mentioned above is allowed. This class of seed used to multiply the certified seed.

Certified Seed: is commercial seed being sold to the general cultivation. This seed need to comply with minimum seed standards as per the Seed Act. 1966 & Rules, 1968.

Truthfully Labelled Seed: Though this class of seed is similar to certified seed but may have minimum standards or even exceed the minimum standards. The seed available in the market belongs to this class and popularity of the varieties or hybrids judged by the quality declared on the label.

Each company or seller may have different seed standards based on the minimum standards. In terms of processing one can classify seed based on its end use. For example if a vegetable seed used like Tomato or Peppers used for commercial seedling distribution then the seed lot should have more than 95 %

germination. For this quality, the processing requirement is different. For export markets this quality may be still more competitive. For ease in understanding and based on processing there may be different seed classes' like-

First Quality (A): Processing of seed for this class must have highly competitive specs or standards in terms of physical purity, germination, genetic purity and free from seed borne diseases. In some crops like pickling cucumbers the seed should be processed to 100% germination. The seed germination standard is declared as 100% on the label for this segment. Also this quality is required for seed used for growing vegetables for processing and for supermarkets packaging where consumers prefer uniform quality. This processing standard is required for some exotic vegetables like broccoli, leek or purple cabbages.

Second Quality (B): Processing for most of the developing countries and for vegetables which are sold in open market and table purpose. A little flexibility in terms of quality is allowed here.

Dirty (C-1): This is a seed processing class for institutional marketing. Sometimes one seller may buy seed from another seller or company to sell under its own brand-in this case a combination of size grading or conditioning will be done and final processing is done after reaching the final seller. This kind of processing is done to minimise costs in terms of treatment, specialized up gradation or blending. This type of seed processing is done mostly for bulk seed movement.

Bulk (C-2 or D): This class of seed processing is for inventory management when the seed is being shipped for specific markets and for storage for some time until the demand for the inventory arises. For example tomato seed required for north India is different from south India. In this case seed grown in south India is primary processed here and shipped to a processing plant in north India where rest of processing is done. Also when demand for a variety is not there and may arise after some time bulk seed after primary size grading or conditioning stored in dehumidified cold storages as bulk. To avoid problems of adverse effect of treatment chemicals or reduce packaging this kind of processing is useful. This type processing is very popular when the seed is imported from other countries like cabbages, cauliflowers or knoll kohl of which seed production is not possible in tropical countries.

1.2 Seed processing of tropical vegetables

Seed processing of tropical vegetables refers to seed production and processing of seed grown in tropical countries like India and also for seed imported for general cultivation though they are not produced here like *brassicas*. Generally some known temperate crops grown under controlled conditions and in hilly regions like Himachal Pradesh, Ooty, Uttarakhand etc.

Indicative list of tropical crops for seed processing in Table 1.

Table 1 : Seed Produced and Imported

Seed Produced in India	Seed Imported for Cultivation In India
Tomato, Capsicum, Hot Pepper or Chillies, Eggplant or Brinjal, Okra, Sweet Corn, Coriander, Onion, Beet Root, Spinach and Other minor leafy Vegetables; Gourds like Watermelon, Cucumber, Squash, Pumpkin, Bitter gourd, Sponge gourd, Bottle gourd, Ridge gourd and other melons.	Processing type Tomato, Radish, Carrot, Cabbage, Cauliflower, Celery, Fennel, Endive, Leek, lettuce, Parsley, Rhubarb, Rosemary, Turnip, Endive and specialized crops like triploid watermelon.

1.3 Flow chart of seed processing

The following chart is on the routine shop floor operations in vegetable seed processing and inter disciplinary approach to ensure quality seeds.

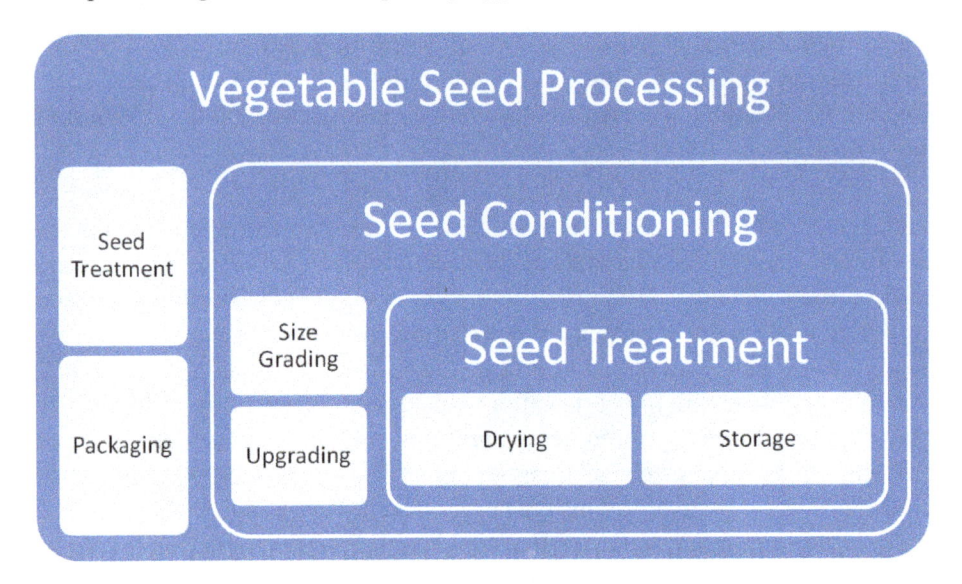

Vegetable Seed Processing
- Seed Treatment
- Packaging
- Seed Conditioning
  - Size Grading
  - Upgrading
  - Seed Treatment
    - Drying
    - Storage

CHAPTER - 2

Harvesting and Seed Pre-conditioning at Farmers Level: Two Case Studies of Tomato and Sponge Gourd

The vegetable seed production in India work on the principle of buy back or contract farming where the production agent or organizer distribute the plots of seed production to growers. The grower is responsible for the rising of seed crop till harvesting or pre-conditioning. The organizer guides the growers regarding the plant protection, roguing and pre-conditioning. The harvesting is entirely a responsibility of the grower but facility for pre-conditioning may be established by the seed organizer. The seed pre-conditioning includes activities like acid treatment on wet seed bases or operations which make seed ready for procurement. After the harvesting or pre-conditioning the seed will be procured by the organizer and head for the processing yard for grading & packaging.

2.1 Harvesting and Pre-conditioning

The methods of harvesting and pre-conditioning various with crop to crop. There are mainly two types of vegetables in terms of type of harvesting.

2.1.1 Wet seeds

This type of seed like tomato, peppers, egg plant, capsicum, watermelon, cucumber, bitter gourd, bottle gourd or pumpkin, where the seed is attached to flesh of the fruit and seed separation will be based on fermentation principle. In this method immediately after harvest of the fruits at physiological maturity allowed to after ripening and seed collected by cutting the fruits. After the cutting the seed along with flesh will be fermented over night in which separation of seed from flesh will takes place. The seed then washed with clean water and gradual drying to optimum moisture will be done. The acid treatment for seed sanitation to ensure free from seed borne pathogens like bacteria, fungi and virus is done on wet seed basis. For this seeds no separate sanitation is required at processing unit.

2.1.2 Dry seeds

This type of crops like sponge gourd, ridge gourd, melons, onion, raddish, carrot, spinach or most of the exotic vegetables. Here the seed is not attached to flesh at the time of harvesting. The flesh holding the seed will be either completely dried or seed bearing on the dry umbels like in onion. This kind of seed is easy to separate but seed sanitation will be on dry seed basis which is done at the processing plant. The seeds of these crops are easy to dry and don't need water much during cleaning.

2.2 Harvesting and Pre-conditioning in tomato (Wet Seed Type)

Maturation of fruits starts at 20-30 days after pollination and harvesting can be done after 50-60 DAP. While picking the fruits one should check the fruit is uniformly red from all sides and also look for the mark of pollination. Such fruits are kept overnight and crushed without using water. The crushed fruit along with pulp and juice kept for fermentation for 24 hours. This will help to wash seed off the pulp easily and also kill seed borne pathogen like bacterial canker. The seed and pulp after fermentation get separated. The seed now washed in running water and dried by spreading a thin layer of mesh for one day in shade and 2 days in mild sunlight during early morning and late evening for 3 times each Min. moisture to maintain is 6%. Always use clean water for washing. Do not expose seeds to direct sunlight since it will affect germination (Fig.2).

Fig.2 : Seed extraction, fermentation and washing in wet seed types

The seed harvested is then subjected to wet seed treatment where acid treatment was done for seed sanitation (Fig.3). Then seed lot is checked for small and infested seeds which are separated by manual hand cleaning (Fig.4).

Fig. 3 : Acid treatment facility for wet and dry seed basis

Fig.4 : Manual primary cleaning before sending to a processing plant

2.3 Harvesting and Pre-conditioning in sponge gourd (Dry Seed Type)

Seed is extracted when fruits become brownish or light brownish stage. At this stage the fruits will become very light. In case of bottle gourd the seeds will be extracted after after-ripening or curing of fruits for 3 days. Here the seeds are extracted by cut opening the fruit when the fruit is half of the original weight, without use of any water. The outer skin of the seeds should be removed by rubbing against hard paper or rubber gloves. The use of water may lead to discoloration of seeds. In case of sponge gourd curing for 4-5 days help in separation of seeds from pulp. By cutting the tip of the fruit seeds can be easily removed and seed dried for optimum moisture. The dry seeds are subjected to seed sanitation as mentioned in tomato.

Fig.5 : Dry seed extraction in sponge gourd

The seed after pre-conditioning will be dried to optimum moisture levels and sent to nearest processing plant in moisture proof bags where these seed lots are stored in de-humidified cold storages before processing & packaging.

Chapter - 3

Processing Line and Equipments in Vegetable Seed Processing

Vegetable seed cleaning and processing are two important aspects of seed post-harvest management. These steps are very essential to maintain the quality of the seed till its next sowing season. The quality of seed produced is dependent on the moisture content and various admixtures. To provide quality seed to farmers, it is very essential to clean seed to its maximum quality. The seed quality attributes like moisture, free water activity, to be controlled in germination, vigor and viability are upgraded during seed cleaning.

Seed cleaning is the term used as a part of processing means separation of admixtures from a particular seed lot. The physical characters like bigger clods, sand, soil, chaffy material, inert matter etc. will be removed during seed cleaning. Seed processing on the other hand is relatively broad term referring to the process of removal of dockage in a seed lot and preparation of seed for marketing is called seed processing. The price and quality of seed is inversely related to dockage, which should not exceed a maximum level permitted for different crops for seed certification. Due to the operation of processing the level of heterogeneity of seed lot gets narrowed down. The heterogeneity occurs in a seed lot due to following reasons:

1. Variability in soil for fertility, physical, chemical and biological properties
2. Variability in management practices (irrigation, application of nutrients etc.)
3. Variability in ability of the seedling for utilizing the inputs
4. Variability in pest and disease infestation
5. Position of pod or fruit in a plant or the position of seed in a pod.

3.1 Principle of seed processing

The processing operations are carried out based on the principle of physical differences found in a seed lot.

Table 2 : Physical properties of seeds and machines used for processing based on them

Physical difference	Suitable machineries
Seed size – varied from small to bold	Air screen cleaner cum grader
Density- ill filled, immature to well matured light weight to dense seed	Specific gravity separator
Shape – round to oval and different shapes	Spiral separator
Surface texture – smooth to wrinkled and rough	Roll mill / dodder mill
Colour of the seed – light colour to dark colour	Electronic colour sorter
Conductivity of seed – low to high	Electronic separator

3.2 Requirement in seed processing

1. There should be complete separation
2. There should be minimum seed loss
3. Upgrading should be possible for any particular quality
4. There should be more efficiency
5. It should have only minimum requirement

3.3 Types of materials removed during seed processing

1. Inert materials
2. Common weed seeds
3. Noxious weed seeds
4. Deteriorated seeds
5. Damaged seeds
6. Other crop seeds
7. Other variety seeds
8. Off-size seeds

3.4 Sequence of operation in seed processing

Sequence of operations is based on characteristics of seed such as shape, size, weight, length, surface structure, colour and moisture content. Because each crop seed possesses particular seed structure, the sequence of operation will be performed with proper equipments. It involves the following stages,

1. Drying
2. Receiving
3. Pre-cleaning
4. Conditioning
5. Cleaning
6. Separating or Upgrading
7. Treating (Drying)
8. Weighting
9. Bagging
10. Storage or Shipping

3.5 The flow chart illustrating the types of materials removed from harvested produce during processing

Harvested seed
↓
Threshed, Shelled and Dried

Inert material
↓
Noxious Weed seed
↓
Deteriorated seed
↓
Damaged seed

Common weed seed
↓
Other crop seed
↓
Other variety seed
↓
Off-size seed

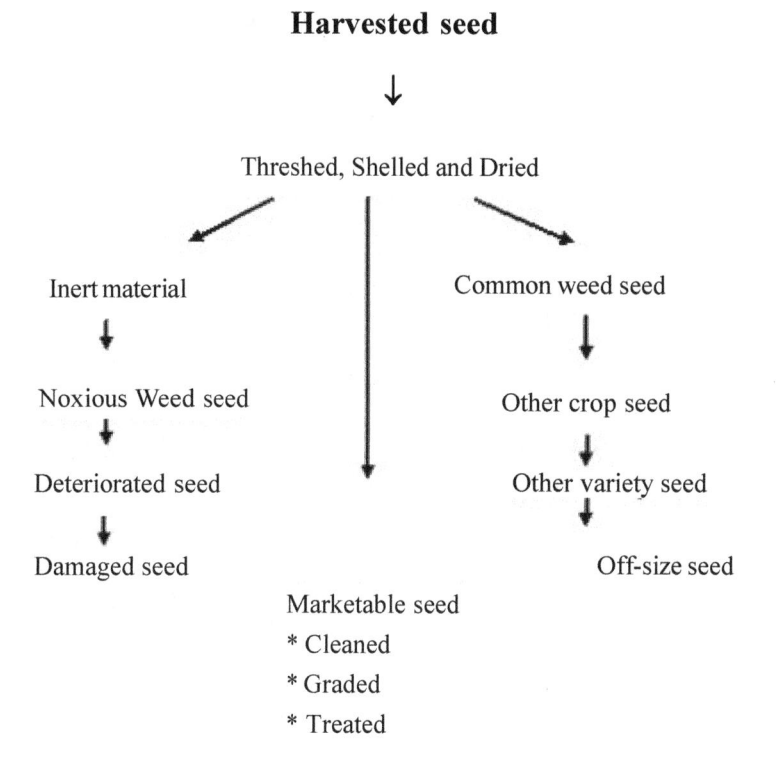

Marketable seed
* Cleaned
* Graded
* Treated

Fig. 6 : The fate of seed from harvesting till processing

Unlike bulky crops, vegetable seed commands a very intensive and closed approach for seed cleaning and processing because of the smaller quantity and high value. We need to employ precision seed sorting techniques to clean vegetable seeds. The admixtures usually make 15-20 % of the total seed lot in vegetables which mainly contain weed seeds, seeds of other crops, light and immature seeds, damaged seeds, larger or smaller seeds, and plant materials, other materials like stones, soil and faeces of rodents. Except some inert material all other materials are hygroscopic and attract moisture. The moisture tend to get transferred to seed. The cleaned seed invite more prices and is basic step in any seed certification programme.

Following are the processing and packaging equipments for shop floor. The seed drying and treatment equipments are covered in respective chapters.

3.6 Seed processing equipments

In any processing plant the assembly of equipments looks interconnected and all the operations from pre-cleaning to packaging are done online or one goes. For vegetable seed processing the basic and must have equipments are air screen cleaner, specific gravity separator and indent disc cylinder. Though the air screen cleaner and specific gravity separator are most widely used. There are two types of processing line based on the volume of seed. The bigger processing line is having an output of 5-10 q / hour where as smaller equipments handle 1-2 q / hr. The working principles of some of the equipments are given here:

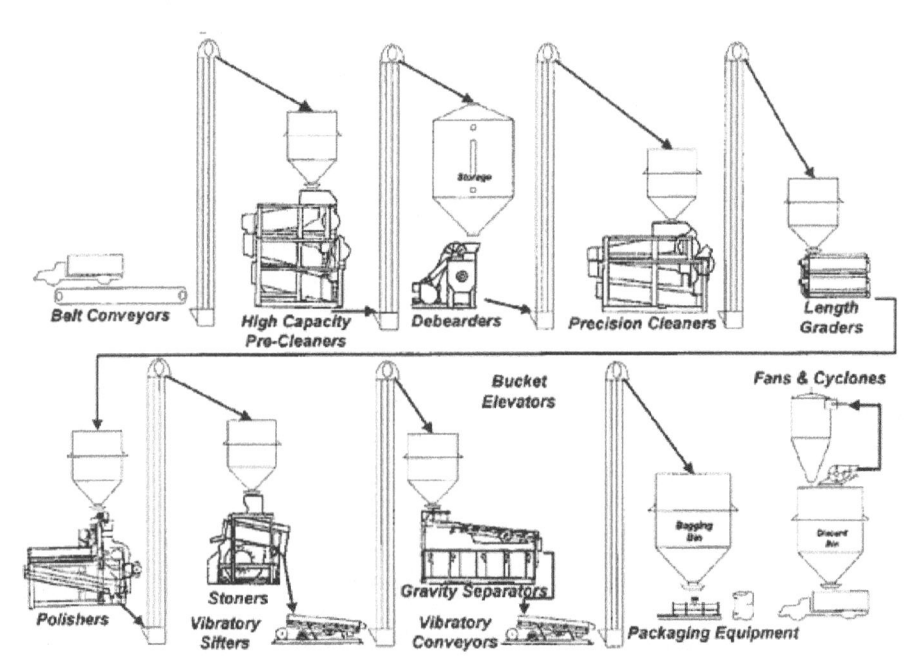

Fig.7 : Shop floor assembly line in a processing plant

Fig. 8 : Processing line for larger volume of seed material

Fig. 9 : Processing line for smaller volume of seed material

3.6.1 Air screen cleaner or pre-cleaner

This type of cleaner is work based on size of the seed. Here the light and inferior quality seeds are get separated from good seed. The product to be cleaned poured into the inlet hopper (1) where the vibratory feeder (2) leads it to the sieveboat. The Triac control (3) sets the capacity. After leaving the vibratory feeder, the product passes the pre-aspiration (4) where light particles are removed. The product is led to the short scalping screen (5) where large particles are removed and led out through chute (6). The overflow continues to the first grading screen (7) where the overflow is led out through chute (8). The through flow continues to the second grading screen (9). The through flow (sand and small particles) from the second grading screen is led out through chute (11). The overflow is led out through outlet (10) and up into the final aspiration channel (12) where the last light particles are removed. These particles settle in the drawer (14) under the aspiration chamber (13). Dust is sucked with the air to the cyclone (15) where the dust particles settle in the glass (16). The prime product leaves in the bottom of the final aspiration channel and settles in the drawer (17).

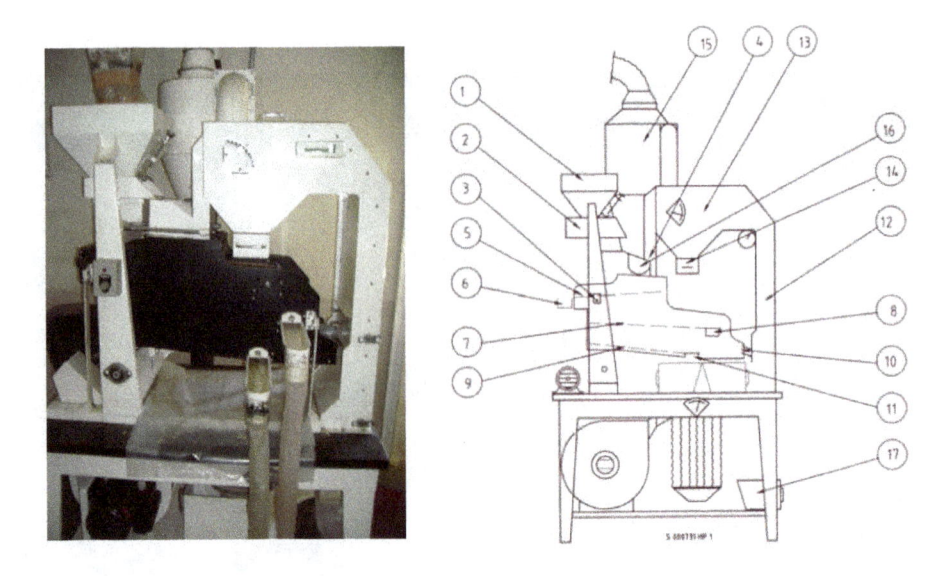

Fig. 10 : Operating procedures of an air-screen cleaner

3.6.2 Gravity separator

This processing machine clean the seed based on density. Here the separation is in phased manner from ill filled, immature to well matured light weight to dense seed. The product to be graded enters through the hopper (1) fitted with adjustable feed gate (2). The motors (3) and (4) are started. The vibrator (5) is started and the material now has to be fed in a steady flow so that it covers the entire table. The separation is analyzed and the necessary adjustments are made.

Adjustment possibilities

    a. Increase or decrease the speed of the eccentric shaft.
    b. Increase or lower the deck lengthways.
    c. Increase or lower the deck widthways.
    d. Increase or reduce the air volume.
    e. Adjust the feed of raw material.

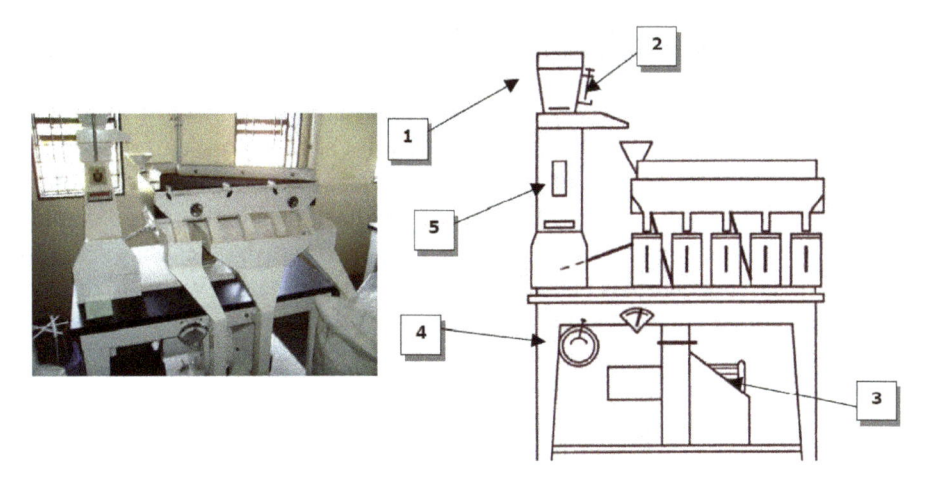

Fig. 11 : Operating procedures of a gravity seperator

3.6.3 Indent disc separator

This type of equipments use length as criteria for seed separation. This machine is designed for continuous flow (or batch) separation where kernels with different lengths but otherwise similar dimensions can be separated from each other. Short seeds are separated from long ones, broken seeds from whole ones, and also round seeds from oblong ones. The machine is fitted with inlet hopper and vibratory feeder with start/stop and potentiometer. Further with speed regulation of the cylinder. The mantle is exchangeable. Motors are for 1 x 240 V, 50 Hz; drive motor is 90 W and motor for vibratory feeder 0.05 kW. The machine has collecting bins and is mounted on a convenient worktable.

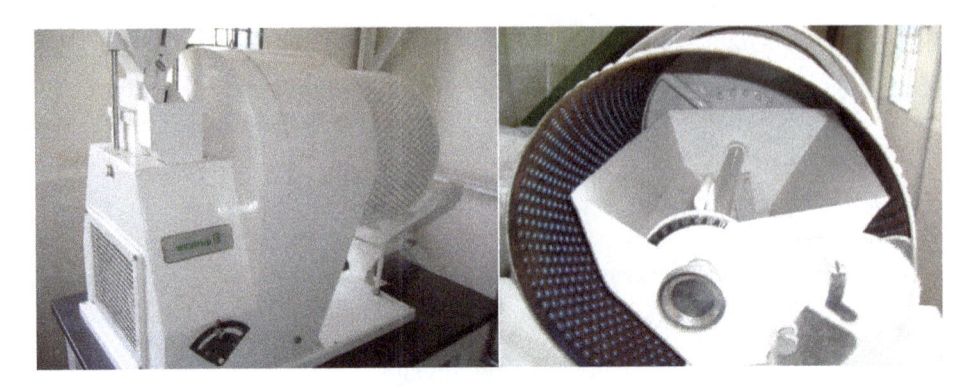

Fig. 12 : Operating procedure of a indent disc cylinder seperator

Other Equipments

3.6.4 Magnetic seed separator

These machines use a magnetic force to separate damaged seeds or other particles from the good seeds. This is used to solve special problems in seed lots that can not be taken out with other basic seed cleaning equipment. Most of the time it is used for vegetable and field seeds that have problems in seed damage (fish mouth in cucumber or melon seeds / cracked radish or cabbage seeds), insect damage (beetle holes in melon seeds) or contamination with soil or plant parts (white plant parts in onion seed) etc.

Fig. 13 : Magnetic seed separator for problematic seeds

3.6.5 Seed mixing/blending machine

These machines are used for mixing/blending of different seed lots to make it a homogeneous lot.

3.6.6 Blower

The use of this as testing machine used for the classification of a sample according to the aerodynamic behaviour (floating speed) of the particles in an air stream The sample is placed in a sample tray with fine mesh in the bottom. Air is blown through the mesh and continues up through the vertical separation pipe. The sample is then divided into two fractions – one in the settling chamber and one in the sample tray (pure seed collection).

Fig. 14 : Mechanical seed blending machine

3.6.7 Feed Hoppers

For all machines the feed hoppers play an important role in easy flow of material and also reduce the labor requirement. Also for bigger machines along with belt conveyors hoppers will work more efficiently. These can be of different capacities based on the type of machine in use and volume of the material.

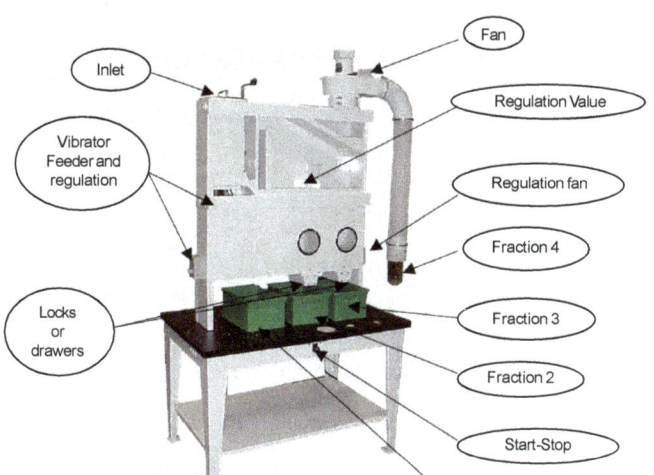

Fig. 15 : Aspirator or seed blower to separate seeds on their aerodynamic properties

Fig. 16 : Feed hoppers for holding and movement of seed in to the processing machines

3.6.8 Screens

For all basic and advanced seed cleaning machines screens are important parts which need to be used based on the type of seed. There will be round (R) or slotted (S) screens and of different sizes measured in mm for crops. Some examples of screen sizes and their types given in Table-2.

(a)

(b)

Fig17a &b : Screen types to separate seeds based on shape

Table 3 : Screen types for processing different crop seed

Crop	Top	Middle	Bottom
Onion	5R	3.5R,3.5R,3.25R	1.75R,1.75R,1.2S
Watermelon	14R	11R	6.5R,6.5R,6.5R
Coriender	7R	4.25R,4.25R,4.25R	1.85S,1.85S,1.85S
Beans	9R	7R,7R,6.85R,	5R, 4.35R,3S
Okra	6R	5R,5R,5R	4.35R,4.25R,4.25R
Bittergourd	14R	11R,11R,11R	6.5R,6.5R,6.5R
Bottlegourd	11r	9.5R,9.5R,9.5R	6.5R,6.5R,4.75R
Ridgegourd	9.5R	7,7,7R	6.4,6.4R,4.75R
Palak	5R	4.25,4.25R,5R	2.5,1.2,1.85R
Watermelon-triploid	6.25R	5,5R,6	1.1,1.1S,4.25R
Cabbage	2.75R	0.9S,1.1S,1.2S	0.9 S
Cauliflower	2.75R	1.5R	0.9S,1.1S,

3.6.9 Seed winnower

The winnower is used to utilise the differences in specific weight between the seeds and other parts. A rough air separation separates the seeds from the husks. It is very suitable for cleaning small seed lots.

Fig. 18 : Seed winnower to clean smaller seed lot

3.6.10 Seed disinfecting or treatment equipment

To eliminate bacterial contamination or seed borne diseases. It can be used as an imbibition unit for the uptake of moisture and for hydration before priming or as a hot water treatment unit. The system uses seed drums to carry the seeds. The seeds are treated in a very uniform and reproducible way.

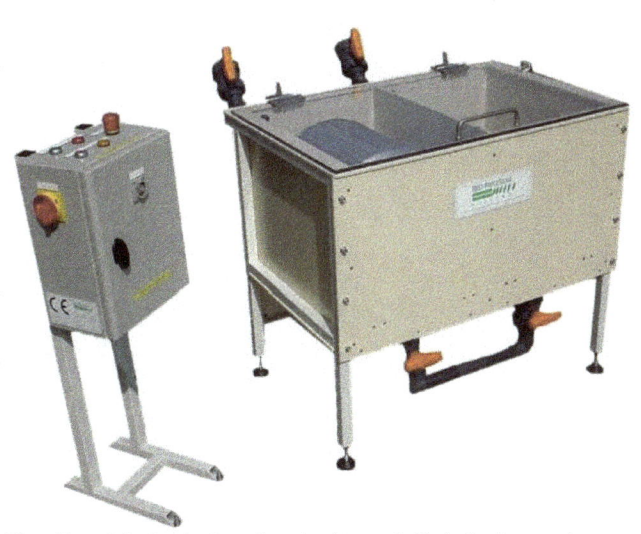

Fig. 19 : Mechanical seed sanitation and dis-infecting equipment

3.6.11 Seed osmotic priming unit

This systems use low water potential osmotica, such as polyethylene glycol (PEG) or slats and air (oxygen) to enhance germination of a seed lot. The goal is to reach a faster and more uniform germination.

Fig. 20 : Osmotic priming unit for seed enhancement

3.6.12 Seed density separation unit

This machine is used for accurate seed density grading using liquids (water or liquids with certain osmotic pressures). During separation, the seeds should keep the same moisture level content.

Fig.21 : Liquid density separation unit

3.6.13 Seed air separator

This air separator is used after the seeds are threshed. By means of air, dust and small plant parts are separated from the seeds. By using this machine a lot of waste material is removed before the fine cleaning process takes place.

Fig.22 : Mechanical seed air separator

3.6.14 Pre-cleaner

This machine is used after the seeds are threshed. This machine separates dust and capsule parts in a fast and easy way, before the fine cleaning process.

Fig.23 : Pre-cleaner to separate chaffy and denser inert matter off seed

3.6.15 Seed head elevator and dosage belt

This elevator is used for transportation and dosing of seed heads or umbels into the inlet of our seed thresher.

Fig. 24 : Seed overhead elevator and dosage belt

3.6.16 Seed extractor

With this seed drill you can manually extract seeds from cucurbita fruit crops.

Fig. 25 : Seed extracting equipment from fleshy fruits (tomato or cucumber)

3.6.17 Seed rinsing channel

This rinsing channel is used to separate heavy seeds from lighter seeds and fruit parts/skins after a seed extraction process.

Fig.26 : Seed rinsing chamber to remove fruit and seed inert matter

3.6.18 Seed washing and sieving unit

This unit is used to separate small quantities of seeds from larger quantities of pulp and skin parts after the seed extraction process.

Fig.27 : Seed washing and sieving unit

3.6.19 Seed centrifuge

This unit is used in combination with our large seed drum. After seed extraction or any kind of wet treatment the centrifuge is used to get rid of moisture that is around the seeds, which makes the drying process that follows shorter.

Fig. 28 : Centrifuge to remove excess moisture off the seed material

3.6.20 Manual and automatic seed treater

These machines used for seed treatment classified based on the working principle. The manual seed treaters are for smaller batches and automatic machines are for bigger batches. The details are covered in chapter on 'Seed Drying'.

Fig.29 : Manual or semi-automatic and automatic seed treaters

3.6.21 Seed dryer

This seed dryer is mainly used for drying back small seed lots. The dryer has 3 compartments and uses conditioned air. This means we dehumidify the air inside the dryer and re-circulate this air. Like this we control the air temperature and relative humidity during the whole drying process, without influence of the ambient air conditions.

Fig.30 : All season controlled seed dryer

3.6.22 Seed sizing machine

This seed calibrating (sizing) machine uses very accurate screens to separate seeds based on their size. You can use round or slotted screens. Calibrating seeds is important if you want go to pack seeds by weight, to create uniformity in young plants and for precision sowing equipment. The machine can be delivered with one, two or three screens. Smaller capacity machines also available, see chapter laboratory seed cleaning equipment

Fig.31 : Seed calibrating or sizing equipment

3.6.23 Seed spiral separator

This machine is used to separate round seeds from other shaped seeds/ parts. E.g. separate ground from cabbage seeds, round from sharp spinach seed and cleavers and very flat seeds from radish. The following weeds are removed from cabbage: sine's grass, triangular shaped seeds, black bind-weed and sorrel.

Fig.32 : Spiral seed grader or separator based on buoyancy and shape of seed

3.6.24 Seed colour sorter

These new generation full colour sorters (Red, Blue, Green) are specially designed for the seed industry. This machine makes a high capacity sorting, based on the colour of the seed. Contamination like discoloured, infected or damaged seeds, soil and weeds can be separated. E.g. white plant parts out of onion, white from black lettuce seeds, green carrot seeds, yellow cabbage seeds etc. These machines have many applications for all kind of seeds, starting from very small seeds like lettuce up to pumpkin.

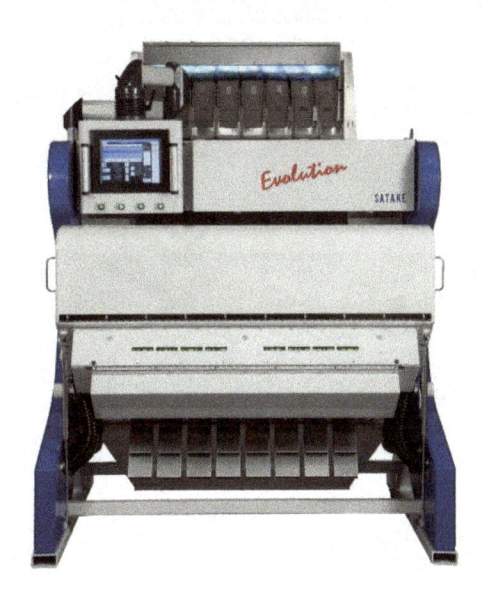

Fig.33 : Electronic colour sorter

3.7 Seed packaging equipments

The packaging of the processed seed is an important activity. Since the size of packaging for vegetable seed is highly variable (from 10 gm to 1.0 kg) it need multitasking or multi capacity packaging machines. Also the packaging machines should be equipped with labeling machines so that to print pouches or tins online. Manual packing can also be done if the packing inventory is small (up to 50 kg).

3.7.1 Mechanical seed packaging

This type of packaging is very useful when the seed material to be packed in smaller quantity like 10, 20, 30 gm. Here the seed has to be weighed and packed simultaneously. Also for very high value seeds which need to be packed in aluminum foils. The seed lot to be packed is fed to the feed bin or hopper. The size of the packing is adjusted in the control box which release only this known quality in to the pouch. The aluminum foil roll is fed at one end or directly pouches will be supplied at pouch supply device. After filling a sealer automatically seal the pouches and pass through the horizontal conveyor. Here with a help of a sensor inkjet printer will print the TL (truthful label) with all the minimum standards on the green surface of the pouch. Also a photo depicting the variety or hybrid will be added on the same machine or manually.

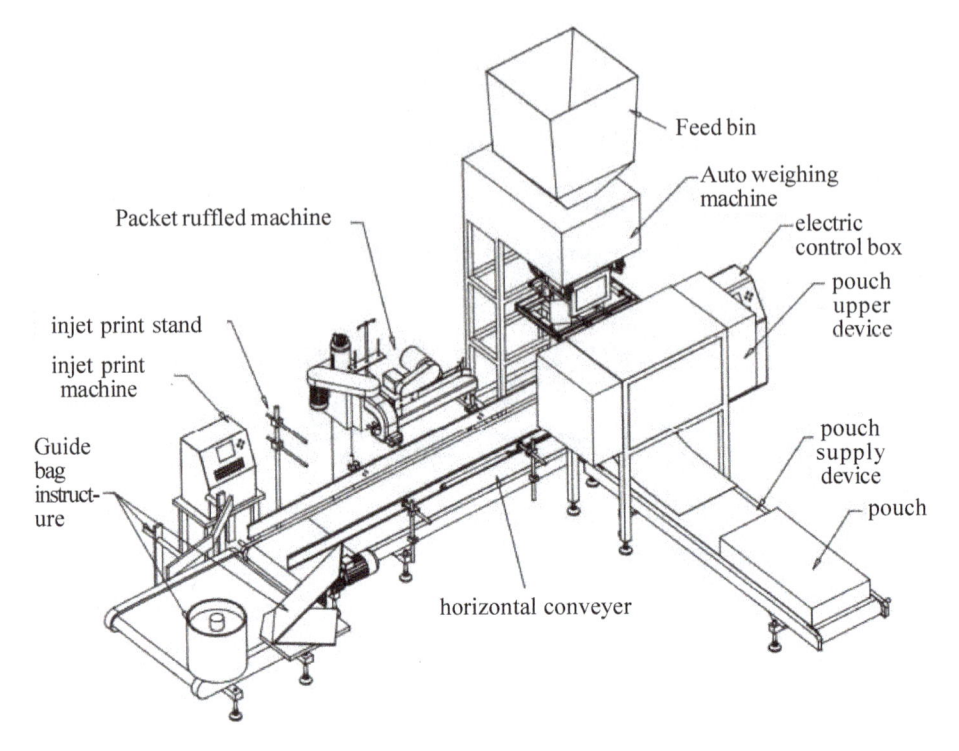

Fig. 34 : Mechanical seed weighing and packing equipment

Here an example of Pakona PK-90 series packing machine is given with technical specifications (complete details can be sought from the company representative).

Fig. 35 : Automatic seed weighing and packing machine (Model Pakona PK-90 Dry)

Table 4 : Technical data of automatic seed weighing and packing models

Technical Data						
Machine Model	PK-90 Simplex	PK-90 Duplex	PK-91 Simplex	PK-91 Duplex	PK-92 Simplex	PK-92 Duplex
Pouch Sizes	W 40 X H 50 mm (min.) W 165 X H 260 mm (max.)	W 55 X H 50 mm (min.) W 80 X H 160 mm (max.)	W 60 X H 50 mm (min.) W 220 X H 260 mm (max.)	W 55 X H 50 mm (min) W 105 X H 210 mm (max.)	W 80 X H 120 mm (min.) W 275 X H 275 mm (max.)	W 55 XH 50 mm (min.) W 140 X H 275 mm (max.)
Speed of the Machine	Up to 80 ppm	Up to 180 ppm	Up to 80 ppm	Up to 150 ppm	Up to 60 ppm	Up to 100 ppm
Roll Width	520 MM	520 MM	660 MM	660 MM	660 MM	660 MM
Power Consumption	4.5 KW	6.0 KW	6.0 KW	7.5 KW	6.0 KW	7.5 KW
Compressed Air(bars)	6	6	6	6	6	6
Packaging Film	All heat sealable laminated films	All heat sealable laminated films	All heat sealable laminated films	All heat sealable laminated films	All heat sealable laminated films	All heat sealable laminated films

3.7.2 Manual seed packaging

Manual packing can be done if the seed inventory to be packed is very small (up to 25-50 kg). The tasks like weighing, packing, sealing and labeling all done manually. It is some time can be labor intensive but useful whenever there is any breakdown or some emergency when seed can be packed immediately.

Fig. 36 : Manual seed weighing, packing and sealing

3.7.3 Personal protective equipments (PPE) in a processing plant

In shop floor there may some tasks which may pose health hazards. Like seed treatment where the persons associated are to handle chemicals. Similarly persons associated with processing have to bear the noise or dust particles. To avoid any health problems to the persons handling seed safety accessories like helmet, goggle, disposable ear plugs, gloves, disposable masks, aprons and safety shoes have to be provided. These safety accessories are available with any industrial supplier and play an important part in implementing ESH (Environment, Safety and Health) practices in any processing plant.

Fig. 37 : Safety at work (various Personal Protective Equipments)

Chapter - 4

Vegetable Seed Processing

The processing of the seed refers to mainly pre-cleaning and sizing. The sizing is up gradation of seed lots based on size, shape or length. In this chapter processing with main reference to sizing or upgrading is given. Sizing or upgrading is done by using different screens in an air screen cleaner, gravity separator, seed blower, indent disc cylinder, belt grader or some time colour sorter. Here specific tasks like use of screens in an air screen cleaner, rubber belt angles when using a belt grader, disc type when using indent disc cylinder or special arrangements in a gravity separator are given. The details of the machines or equipments with the principles of working were already covered in Chapter-3.

4.1 BEET ROOT

Pre - cleaner:

Upperscreen : 12.00 mm round
Bottomscreen : 1.75 mm slot-screen and/or 2.40 mm round

Air screen cleaner:

Large variety:

8.50 - 9.00 mm round screen

7.50 - 8.00 mm round screen

2.00 mm slot-screen and/of 2.60 - 3.00 mm round screen

Small variety:

7.00 – 8.00 mm round screen

6.00 - 6.50 mm round screen

1.70 mm slot-screen and/of 2.60 - 3.00 mm round

Belt grader: To remove sticks attached to the seeds. Rough rubber belt.

Angle : +/-27°

Belt speed : **+/-12.5**

Motor setting : 20

Chute : +/- 8

Cap./hour : +/- 25 Kg/hour.

Gravity separator: Wire-netting - to remove pieces of Barley.

Indented cylinder: Cylinder 8.00 mm round – to remove foreign material that is longer in length.

4.2 BRASSICAS (Cabbage, Cauliflower, Knol Khol)

Source : Peter M. Driuk

Air screen cleaner : The set up, regarding screens, depends on the type/model A/S cleaner used.

Upper screen : 2.60 / 2.75 mm Round

Middle screen : 2.25 / 2.50 mm Round

Bottom screen : 1.00 / 1.20 mm slot

Vegetable Seed Processing

Spiral Separator: Remove foreign material (chaff, plant debris, weeds, soil particles etc.); and try to minimise seed loss as much as possible. Please use the "fingers", by doing this the loss will increase and because of that the original clean out needs to be remilled until no "good" seeds are left in the clean out.

Gravity separator: Remove foreign material (chaff, plant debris, weeds, soil particles, etc.) and +/- 1% of the most light weighed seed.

Colour sorter: Use of colour sorter if weeds of gallium are noticed in physical purity test.

4.3 CARROT

Pre-milling : Use the De-Awner only for seed that has not been debearded.

Upper screen   : 3.00 mm Round
Bottom screen   : 1.00 mm Round

Air screen cleaner: With De-Awner

- When seed is pre-milled or combine harvested, process the seed with the screen selection from the second process run.
- Usually the De-Awner setting is adjusted on the second process run.

First Process Run: 2.75 / 2.40 Round   1.30 slot   1.00 Round.
Second Process Run: 1.10 / 0.90 Slot   2.40 / 2.00 Round   0.80 / 0.90 Round

Indented cylinder

Recommend using an Indented cylinder to remove undesirable material that is longer or shorter in length. This could be foreign material (sticks, half or short seeds, plant debris, etc.) or seed mixtures (cucumber, melon, etc.).

Cylinder: 4.50 to remove large parts.

Cylinder: 2.50 to remove small parts.

Gravity separator

Remove foreign material (chaff, plant debris, weeds, dirt clods, etc.) and light seed.

4.4 CARAWAY

Air Screen cleaner

Upper screen	: 3.30 mm round-screen
Middel screen	: 1.60 mm slot-screen
Bottom screen	: 0.80 mm mess-screen

Indented cylinder

Cylinder 3.00 mm round to remove small weeds.

Cylinder 6.00 mm round to remove Lolium (longer than the seeds).

Screener

- 1.50 mm slot-screen to remove Galium and Atriplex

Spiral separator : To remove cabbage

4.5 CELERY

Air screen cleaner

    Upper box   : 1.50 mm round   0.50 mm mess-screen

    Bottom box : 0.80 mm slot-screen   0.60 mm mess-screen

Indented cylinder

    Cylinder 2.50 - 3.00 mm round   to remove foreign material that is longer in length.

Gravity separator

    To remove plant debris (chaff and sticks)

Remarks

    Separation of doubled seeds is no option, because soon (6 months) after the rubbing/polishing there is often a decrease in germ (30-50%).

4.6 CHICORY

Air screen cleaner

Upper box: 2.40 mm round   0.90 / 1.00 mm round
Bottom box: 2.00 mm round/1.40 mm slotscreen   0.70 / 0.80 mm slotscreen

Indented cylinder (big)

    Cylinder 2.75/3.00 mm round to remove foreign material that is shorter in length.

    Cylinder 4.50 mm round to remove foreign material that is longer in length.

Indented cylinder (small)

    Cylinder 2.25 mm round to remove foreign material that is shorter in length.

    Cylinder 3.50 mm round to remove foreign material that is longer in length.

    Swing: To remove sticks and chaff

    Gravity separator: To remove chaff and soil parts.

4.7 CORIANDER

Indented cylinder is always the first process step (because of the presence of damaged seeds).

1. Indented cylinder

Cylinder 6.00 mm round to remove foreign material(Avena, Grain, Vicia Spp. and Sclerotien) longer in length.

2. Air screen cleaner

- ***Small seed variety***

    Upper box :   > 4.50 mm round-screen

    < 1.80 mm slot-screen (damaged seeds)

    Bottom box : > 4.00 mm round-screen

    < 0.90 mm slot-screen } (chaff)

    < 2.10 mm triangle-screen } To remove Polygonum Convolvulus

- ***Large seed variety***

    Upper box :   > 6.00 mm round-screen

    < 2.50 / 2.75 mm slot-screen (damaged seeds)

    Bottom box : > 4.75 mm round-screen

    < 2.10 mm triangle-screen

3. Gravity separator

Not a standard process step.

4.8 SWEET CORN

Pre - cleaner

    Upper screen   : 4.75 mm round-screen
    Bottom screen   : 1.20 mm slot-screen

Air screen cleaner

    Upper box : 4.50 mm round-screen   1.50 / 1.60 mm round-screen.
    Bottom box : 4.25 mm round-screen   1.20 mm slot-screen
    Gravity separator: To remove chaff.

4.9 GHERKINS OR PICKLING CUCUMBER

Air screen cleaner

Only Open pollinated !

Upper screen	: 4.80 - 5.40 mm round
Middle screen	: 1.80 - 2.30 mm slot-screen
Bottom screen	: 3.10 - 3.30 mm round

Remarks: Hybrid varieties should never be processed on the Air screen cleaner, because the seeds can be damaged.

Swing: To remove sticks and seeds with fruit skins attached to it and deformed seeds.

Screener: > 1.70 mm slot-screen to remove deformed seeds.

Indented cylinder: Cylinder 7.00 to remove broken/damaged seeds.

Gravity separator: With wire netting to remove deformed seeds.

Air colomn separator: To remove blind seeds.

Electronic color separator: To remove Sclerotien and discolored seeds.

4.10 CUCUMBER (GENERAL)

1. Cleaning

a. Hand cleaning	:	to remove dark fruit skins attached to the seed.
b. Air Column Separator	:	to remove chaff and empty and light seeds.
c. Indented cylinder	:	cylinder 8.00 to remove half and deformed seeds

2. Sizing for length

Using indent cylinder 10.0 mm

Two length sizes are created:

8.00 - 10.00 mm   6.50 – 8.00 mm (for small seeded varieties)

> 10.00 mm   > 8.00 mm (for small seeded varieties)

These length sizes become 2 separate batches and will be sized for thickness on the screener.

3. Sizing on width and thickness

a. On width

- Remove seeds < 3.50 or < 4.00 mm round (varies per batch). This size contains a lot of deformed seeds.
- The supervisor reviews each batch and decides whether a batch will be discarded or saved.

b. On thickness using Slot screens

First Process Run: 1.40 slot. 1.50 slot. 1.80 slot.

Second Process Run: depends on the quantity:

< 1.40 slot. Repeat on 1.30 slot.

> 1.80 slot. Repeat on 1.90 slot.

* Screen sequence in machine: 3.25/3.50 R, 1.40 S, 1.50 S, 1.80 S (first process run)

Remarks referring to sizing: Screen size < 1.30 mm slot:

4.11 PUMPKIN

Vegetable Seed Processing 41

Air screen cleaner

    Upper box   : 12.00 mm round   4.00 mm round

    Bottom box   : 9.00 mm round   4.50 mm round

    Swing: To remove shell parts, seeds with shell parts attached and chaff.

    Screener :

        > 3.20 mm slot-screen } rejects

        < 2.20 mm slot-screen } rejects

4.12 CUMIN

Air screen cleaner

    Upper box   : 5.00 mm round   1.00 mm mess-screen.

    Bottom box   : 1.50 mm slot-screen   0.70 mm slot-screen.

Indented cylinder

    Cylinder 4.50 mm round to remove chaff.

    Cylinder 5.00 mm to remove Plantago.

    Screener: 2.10 mm round to remove Lolium.

    Gravity separator: To remove sticks.

4.13 DILL

Air screen cleaner

    Upper screen   : 3.00 / 3.25 mm round-screen
    Middle screen : 1.20 / 1.30 mm slot-screen
    Bottom screen : 1.30 mm round or 0.90 mess-screen

Indented cylinder

    Cylinder 2.75 or 3.0 mm round to remove chaff and/or weeds shorter in length

    Cylinder 5.50 or 6.00 mm round to remove foreign material longer in length.

    Carter indent: To remove chaff and/or weeds (Solanum Nigrum, Chenopodium, Sinapis Arvensis and Cuscuta Europaea) shorter in length.

    Screener

    > 0.90 mm slot-screen to remove chaff and Rumex.

    < 1.25 mm mess-screen to remove Lolium.

    Indented cylinder: Cylinder 3.25 mm to remove weeds shorter in length.

    Gravity separator: To remove Linum Usitatissimum (Fig.5).

4.14 EGGPLANT OR BRINJAL

Air screen cleaner

    Upper box: 2.00 mm slot-screen   2.00 mm round

    Bottom box: 1.40 / 1.50 mm slot-screen   2.10 mm round

Screener

    Upper screen   1.20 - 1.40 mm slot-screen to remove doubled and swollen seeds.

    Bottom screen   2.00 mm round.

Air colomn

    To remove blind seeds.

Color sorter

    To remove discolored seeds.

    Sizing : (only when requested by inventory management)

    Size in increments of 0.20 mm Round between 1.80 mm and 3.20 mm

4.15 FENNEL

Air screen cleaner

    Upper box   : 4.00 mm round-screen   1.60 / 1.80 mm round-screen
    Bottom box : 1.60 mm slot   0.80 mm slot-screen

Indented cylinder

    Cylinder 4.00 mm round to remove chaff and or weed shorter in length.

Gravity separator

    To remove chaff, dirt clods and insect damaged seeds if present.
    Air column: To remove small and insect damaged seeds.
    Color sorter: To remove dark seeds
    Sizing (on length)
    4.50 – 6.00 mm
        > 6.00 mm

Vegetable Seed Processing 45

4.16 FENUGREEK

Air screen cleaner: 3.75 mm round   1.25 mm slot-screen

Indented cylinder

  Cylinder 3.25 mm round to remove chaff shorter in length.
  Cylinder 8.00 mm round to remove grain longer in length.
  Spiral separator: To remove Galium

4.17 ENDIVE

Air screen cleaner

  Upper box: 2.25 or 2.40 mm round   0.90 mm round screen
  Bottom box: 1.40 mm slot-screen   0.70 mm slot-screen

Screener: (only for pre-pelleting quality)

Upper screen: 1.65 mm rond   > 1.65 mm round

Bottom screen ± 0.83 mm slotscreen   < 0.83 mm slotscreen

The seeds between 0.83 mm slot – 1.65 mm round screen is pre- pelleting quality.

Screener (solutions for weed problems).

Polygonum convolvulus	>1.40 mm slot-screen rejects
Setaria	< 1.20 mm round screen rejects
Polygonum Percicaria (small)	< 0.88 mm slot-screen rejects
Polygonum Percicaria (coarse)	> 1.60 mm round rejects
Solanum Nigrum	< 0.70 slot-screen rejects
Picris Echioides	< 0.88 slot-screen rejects
Lettuce	< 0.88 slot-screen rejects
Echinogloa	> 1.50 mm round rejects

Indented cylinder

Pre-pelleting quality	< cylinder 2.75 mm to normal quality.
Normal quality (cleaning)	**< cylinder 2.50 mm rejects.**

Electronic color separator

To remove sclerotien and dark coloured chaff.

Gravity separator

Pre-pelleting quality	- To remove soil and chaff (to normal quality)
Normal quality	- To remove soil and chaff (rejects)

4.18 LEEK

Air screen cleaner

Upper box	: 3.10 mm round-screen	1.00 mm slot-screen
Bottom box	: 2.80 / 2.90 mm round	1.00 / 1.20 mm slot-screen

Gravity separator

To remove chaff and dirt clods.

Indented cylinder

Cylinder 2.00 mm round	to remove foreign material that is shorter in length.
Cylinder 3.50 mm round	to remove foreign material that is longer in length.

Electronic color separator

To remove all of the light colored chaff and almost all kinds of weed.

Screener

< 1.00 Slot-screen	to remove Trifolium Picris Setaria.
> 1.70 Slot-screen	to remove Polygonum Convolvulus, Geranium Dissectum

4.19 LETTUCE

48   Vegetable Seed Processing

Air screen cleaner

    Upper box: 0.90 / 1.002.70 mm Gauze   0.80 mm round.
    Bottom box: 0.80 / 0.90 mm slot-screen   0.90 mm round.

Indented cylinder: Cylinder 3.25-3.50mm round to remove chaff, weeds and short seeds.

Gravity separator: To remove chaff, sticks and empty seeds

Electronic color separator: To remove variety mixtures (only for pelleting quality) It can also be used to remove chaff and latex.

4.20 WATERMELON

Indented cylinder: Cylinder 7.0 - 10.0 mm Cylinder selection depends on the *seed length*.

Example: Use of a 10.0 mm size cylinder :- First process run put the inside trough in the low position. Repeat the process and let the seed larger than 10.0 become a separate batch. Process one to two times the seed smaller than 10.0 with the inside trough in the upward position. This process step removes halves, broken and deformed seeds then put the inside trough in the low position and reprocess the seed smaller than 10.0 three to four times. This will separate the long seeds from the short seeds. Finally add the long seeds to the batch that is larger than 10.0 mm in length.

Screener : Larger than 10.0

Vegetable Seed Processing 49

Upper screen: Use between / 1.90-3.00 mm slot to remove deformed seeds.

Bottom screen: Use between / 3.00-4.50 mm round to remove small seeds.

Smaller than 10.0

Upper screen: Use between / 1.70-2.80 mm slot to remove deformed seeds.

Bottom screen: Use between / 3.00-4.50 mm round to remove small seeds.

Blower

Used to remove empty and light seeds.

Process each batch separately (larger/smaller than 10.0) on the Blower.

First Process Run: *Remove empty seeds.*

Second Process Run: Remove *5% light seeds.*

Third Process Run: Remove another 3% light seeds.

4.21 OKRA

Air screen cleaner

Upper screen	: 5.20 mm round
Bottom screen	: 3.00 mm round screen (3.25 mm slot screen only to broken seeds)

4.22 ONION

Air screen cleaner

    2.10 Triangle/ 3.50 Round   1.60 / 1.75 Round
    3.00 Round   1.10 / 1.20 Slot

Indented cylinder

Recommend using an Indented cylinder to remove undesirable material that is longer or shorter in length. This could be foreign material (sticks, plant parts, etc.) or seed mixtures (cucumber, corn, etc.).

Gravity separator

Remove foreign material (chaff, plant debris, weeds, dirt clods, etc.) and light seed.

Electronic Color Separator: To remove light coloured plant parts, weed seeds and seed mixtures.

4.23 PARSLEY

Vegetable Seed Processing 51

Air screen cleaner

Upper box	: 2.40 mm round	0.90 / 1.00 mm round
Bottom box	: 2.20 \ 2.30 mm round	0.80 mm slot-screen

Indented cylinder

Cylinder : 1.75 – 2.00   To remove foreign material or weeds shorter in length

Cylinder : 4.00   To remove foreign material or weeds longer in length

Gravity separator: To remove foreign material (chaff, sticks).

4.24 PEPPER (Hot Pepper or Chillies and Capsicum)

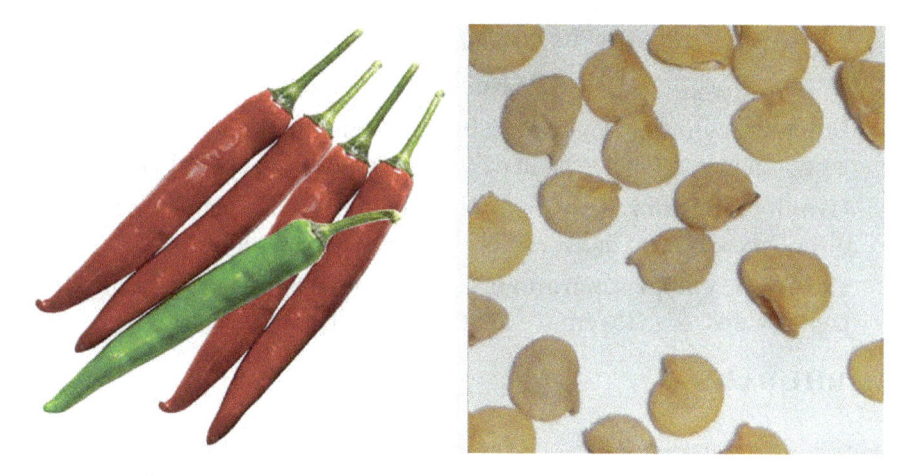

Air column separator: Remove light weighed seeds.

Electronic color separator: Remove discoloured /black seeds.

Screener: - Remove double, deformed seeds and fishmouths (Open Hilum): Screen Recommendations: >1.50-1.90 mm slot (varies by variety)

Remove thin seeds: Screen Recommendations: < 0.90 mm slot (+/- 3%)

Screen-sizes

2.30 - 2.50 mm Round
2.50 - 3.00 mm Round
3.00 - 3.50 mm Round
3.50 - 4.00 mm Round

4.25 RADISH

| Air screen cleaner: | 4.00 Round. | 1.20 Slot. |
| | 3.75 Round. | 1.2 / 1.4 Slot. |

Indented cylinder: Cylinder 6.0 mm to remove foreign material that is longer in length (sticks, plant debris, dirt clods etc.) or seed mixtures.

Gravity separator: Remove foreign material (chaff, plant debris, weeds, dirt clods, etc.) and light seed. Use often (wire) netting.

Electronic colour separator: Used to remove Sclerosis and Polygonum convolvulus. Also to remove light coloured damaged seeds.

4.26 RHUBARB

Air screen cleaner

| Upper box | : 11.00 mm round | 2.40 mm slot-screen |
| Bottom box | : 9.00 mm round | 4.50 mm round |

Vegetable Seed Processing

Gravity separator

To remove chaff and sticks.

Swing: To remove chaff.

4.27 ROSEMARY

Indented cylinder: Cylinder 3.50 mm to remove foreign material longer in length.

Gravity separator: To remove chaff and blind seeds.

Belt grader: As an option to remove sticks.

4.28 RUE

Air screen cleaner

    Upper box   : 1.75 mm round-screen   1.25 mm round-screen
    Bottom box : 1.30 mm slot-screen   0.80 mm slot-screen

Indented cylinder

    Cylinder 2.00 to remove chaff.
    Cylinder 3.50 to remove Lolium.

Spiral separator

    To remove chaff.

Debearder: To remove dirt clods (45 minutes).

Gravity separator: To remove chaff and or dirt clods.

4.29 SPINACH

Pre-Cleaner: This process step is used for very dirty batches without the use of a De-Awner.

    Upper screen   : 7.00 R.
    Bottom screen : 2.00 R.

Air screen cleaner: This process step is used in combination with a De-Awner. Use the De-Awner carefully with weak varieties!

    Upper box   : 6.00 R./5.00 R.   2.00 R.
    Bottom box : 4.50 R.   2.25 R./ 2.30 R.

Remarks in connection with screen-choice: Sometimes a slot-screen is used as a bottom screen to remove weeds. Use a triangle screen 1.90 /2.40 mm instead of a round screen when a batch contains Polygonum Convolvulus. When the batch is mixed with **peas and/or wheat,** first use the Air Screen cleaner to

Vegetable Seed Processing

remove them. After that use the De-Awner to remove chaff and soil particles.

Gravity separator: Remove foreign material (chaff, plant debris, weeds, dirt clods, etc.) and light seed. Use the (wire) netting to remove pea-parts, stones and wheat (fr5).

Indented cylinder

Cylinder is used to remove large parts of chaff and sticks (if necessary). Cylinder 6.00 / 7.00

4.30 SQUASH OR ZUCCHINI

Check batch for attached seed skin parts : Place 10 minutes in rotating

Air screen cleaner

10.00	-13.00 Round.	5.00 – 6.00 Round
4.25	- 4.50 Slotted	1.80 - 2.50 Slotted

> 4.25 Slot screen optional for open/deformed seeds.

Indented cylinder: Cylinder 14.00   removal of half seeds.
Swing: Removal of chaff parts and loose skinparts.

Mess   : 3.
Cap/hour : 300.
Screener:
4.25 mm slot removal of open/deformed seeds.
7.50 mm Round removal of small seeds.
Air column: Removal of light seeds.

4.31 TOMATO

Air screen cleaner: Remove foreign material (chaff, plant debris, weeds, dirt clods, etc.) and light seed.

Rough/Non-Debearded Seed   4.00 Round (upper screen)

                                        1.70 - 2.10 Round   (Bottom screen)

Smooth/Debearded Seed   3.20 Round (upper screen)

                                        1.40 - 1.60 Round   (Bottom screen)

Indented cylinder

Cylinder 7.00 mm - to remove foreign material that is longer in length (such as sticks)

4.32 TURNIP

Air screen cleaner

    Upper box   : 2.25 mm round   0.80 mm slot-screen.
    Bottom box   : 1.85 / 2.00 mm round   0.90 mm slot-screen.

Indented cylinder : Cylinder 3.00 mm to remove Lolium.

Spiral separator: To remove Lolium and chaff.

Screener:> 1.14 mm slot-screen to remove Amaranthus/Chenopodium.

Table 5 : Seed processing by using machines to remove weed seeds:

Weeds/mixtures	Machine	Information
Amaranthus and/or chenopodium	Indented cylinder (long) ∏ Indented cylinder (50 cm) ∏ Screener →	Cylinder 2.50 Cylinder 2.00 / 2.25 0.70 mm slot
Cuscuta	Long Indented cylinder (carter) → Magnetic separator →	Very good result, but with a large loss! (cap. 40,- kg/hour). In the USA good results, No experience in Enkhuizen yet.
Echinochloa	Screener Magnetic separator → Gravity separator →	< 1.80 round - > 0.90 slot rejects > 1.80 round - > 1.05 slot rejects No result No result
Lolium	Indented cylinder (long) ∏ Indented cylinder (50 cm) ∏	Cylinder 4.5 Cylinder 3.50 / 4.00
Picris Echioides	Screener → Elec.colour separator →	< 0.80 netting or 1.10 round – rejects 75,- kg/hour
Polygonum persicaria	Screener →	Possibly 1.10 / 1.20 slotOr after sizing per fraction on the Indented cylinder
Polygonum aviculaire	Screener → Long Indented cylinder →	> 1.00 slot – rejects-(90% clean seed) Cylinder 4 (+/- 40,-kg/hour)
Setaria Viridis	Screener →	< 1.20/1.30 R– rejects (regain sometimes rejects on slot-screen)
Solanum Nigrum	Indented cylinder (long) ∏ Indented cylinder (50 cm) ∏ Screener → Gravity separator →	Cylinder 2.5 / 3 Cylinder 2.50 (regain sometimes rejects on slot-screen) < 0.80 slot – rejects no results
Veronica agrestis	Indented cylinder ∏ Long Indented cylinder →	Cylinder 2.00 Cylinder 4 (40,-kg/hour)
Savory Summer	Indented cylinder ∏	Cylinder 1.75
Centaurea	Indented cylinder ∏	Cylinder 3.50
Cheiranthus	Indented cylinder ∏	Cylinder 2.50
Cabbage	Screener →	1.00 slot
Leek	Screener →	> 1.25 (wire) netting
Lettuce (white)	Elec. Colour separator ∏	75,- kg/hour

CHAPTER - 5

Seed Processing for Improving Quality and Purity

Seed processing is specialized activity and also a tool to improve seed quality further to an acceptable standard. Seed quality parameters like germination, vigour or genetic purity can be improved using processing. Say a seed lot having a minimum standard of 90% germination, but the competition in the market offering seed quality 92 or 95 % then to meet the market demand we can process the seed lots using re-milling or re-conditioning techniques. Also sometimes simple techniques like liquid density separation which uses specific density of the seed to separate good seed from inferior quality can be considered. Various case studies on processing to improve seed quality are given here.

5.1 Processing to improve seed genetic purity

Seed sorting is one of the important operations of a seed processing and conditioning plant. The separation of good seeds using appropriate screens is an effective method to ensure high yield potential and quality. This operation is the basic post-harvest operation for any seed crop and a pre-requisite for successful marketing (Agrawal, 1996). The efficiency of processing determines the performance of the hybrid in the market as well as financial returns to the seed grower. Seed processing ensures the highest seed quality is maintained in the storage and during marketing. Though the precision sorting to improve germination is debatable fact, but to improve hybridity or genetic purity of seed is yet to be known.

Although hybridity is biological and cannot be physically increased, it can be used effectively for sorting seeds where there is a significant difference in seed size between the parental lines and the hybrid. In eggplant (*Solanum melongena* L.) hybrid seed production (No.1461734) seed sizes of the parental inbred lines and the hybrid vary significantly. In this eggplant hybrid due to profuse flowering character of female parent, there is a high likelihood of selfing. Ideally, selfed flowers should be removed from the female parent before developing into selfed fruits and get mixed up with the crossed ones. As it is difficult to identify selfed fruits at the time of harvesting, these selfed fruits are harvested along with crossed ones resulting in mixed seed lots of low genetic purity. Because of the high cost of hybrid eggplant seeds, there is a need to develop efficient methods for separating selfed seed from hybrid seed. The seed lot after sorting can be tested for genetic purity and verified with the results of the test before seed sorting.

So in the similar lines an attempt was made at Department of Seed Science and Technology, UAS, Raichur, for improving the hybridity of a popular hybrid of a leading company by precision sorting method. This particular seed lot of the hybrid (No.1461734) was declared having low genetic purity and germination than the required standard. To classify smaller seeds, a physical purity test (ISTA, 2005) was performed and seed sizes were compared with guard samples of both the parents. The percentage of smaller seed was quantified based on this test and smaller seeds were removed by using the smaller screen (<2.0 mm) than the recommended size for this hybrid.

A single 33.4 kg seed lot of hybrid eggplant (No. 1461734) was used for this study. The results of genetic purity and germination tests before starting the experiment show genetic purity of 80.4% and germination of 59% which were considered as the control. The seed size of the male parent (2.0 to 2.2 mm) and the hybrid are almost similar (2.0 to 2.3 mm) but the seed sizes of the female parent were smaller than (1.8-2.0 mm) the hybrid. The physical purity test showed there is 20 % smaller seed, which is about 6.68 kg of the total lot. This indicates clearly that there is strong possibility of the admixture of selfs instead of crossed seed. The treatments for screen aperture size were imposed by keeping the top and middle sieves as constant sizes of 4.0 (round) and 3.5 mm (round), respectively, and the bottom screen aperture was changed by increments of 0.1 mm from 1.7 to 2.1mm (round), and in increments of 0.1 mm from 0.8 to 1.2 mm (slotted).

To avoid any errors in the results both genetic purity and germination tests were performed again before starting the experiment. The physical character of the seed such as seed size was analyzed for the hybrid seed lot, seed sample of male parent and female parent by using the MARVIN digital seed analyzer (GTA Sensorik GmbH, Germany).

The seed lot was then sorted using an Agrosaw vegetable seed air screen cleaner with 3 long screen layers for continuous flow separation. The seed retained on each of the sieves were quantified. Seed quality parameters, namely hybridity %, germination % were quantified for each lot using ISTA (2011) seed testing procedures. The hybridity or genetic purity tested by plant grow-out test in the field. The percentage of usable transplants was quantified by sowing seeds in pro-trays in the nursery which was established in the green house. The usable transplant percentage is healthy, well grown, saleable seedlings or transplants obtained per gram of seed after 10-14 d of sowing. The germination percentage was determined by roll towel method.

Then the data was statistically analyzed Random Block Design following the method of Panse & Sukhatme (1985) to test the significance of treatments and to draw conclusions.

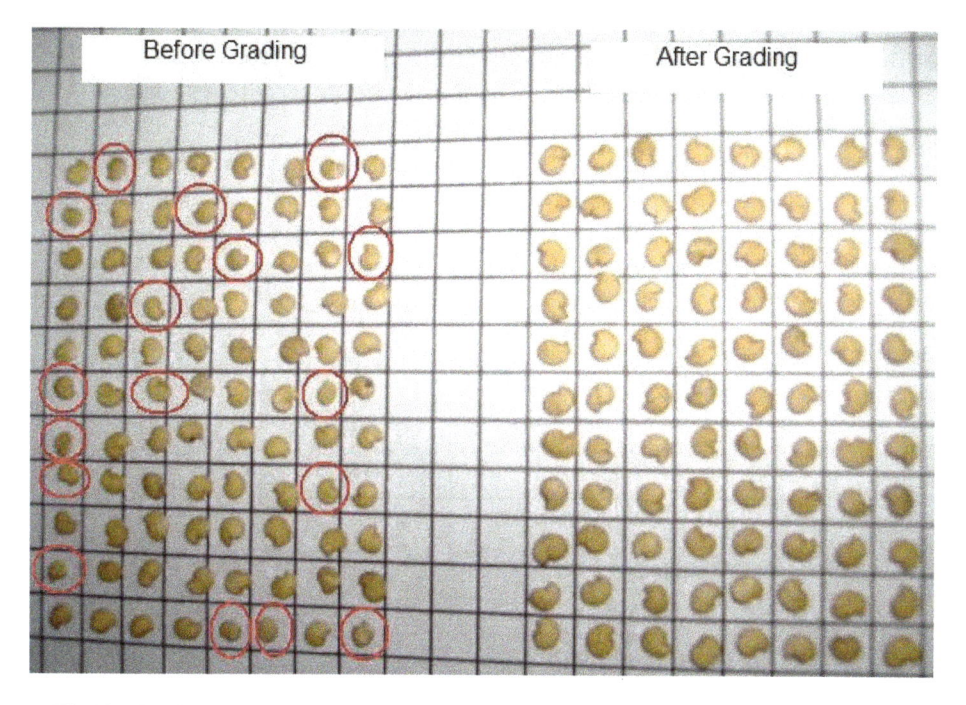

Fig.38 : Seed types before and after precision sorting to differentiate seeds of inbred lines

The results of the study showed that the seed sorted using a bottom screen of 0.9 mm (slotted) resulted in higher hybridity (95.57%) in the PGO test, followed by a bottom screen of 0.8 mm (slotted) (Table 1). The hybridity percentage in seed sorted using round bottom screens, ranged between 82.93 to 85.07 %, which is a little improvement over the control. The improvement in the hybridity was not affected by the round screen apertures, where as the slotted screen apertures significantly improved the hybridity. The slotted screen (0.9 mm) though improved the germination as compared to the control, it is the round screen (2.0mm) which was recorded highest germination (79 %) and recovery of usable transplants (76%).

A positive association between seed size and seed quality parameters was reported by Srimathi and Vanangamudi (1993) in cowpea and Kalavathi and Vanangamudi (1990) in clusterbeans. The study also support the fact that the seed sorting can be very useful in improving certain seed quality parameters such as germination by precision seed sorting, as also reported by (Hanumaiah and Andrews, 1973) and the bigger seed size directly influences the seed quality both in field and storage (Pandita and Randhawa, 1995). This study also support the observation by Komba et al. (2007) on effect of seed size within a seed lot on germination.

Though the study confirmed the positive association between seed size and germination or recovery of usable transplants, there were no prior findings on this aspect but this study offers a practical solution in hybrid vegetable seed production involving inbred lines which produce different seed sizes. The possibility of improving hybridity using precision seed sorting can be used for blending of smaller seed lots with similar hybridity and to reduce the number of batches. Reduction in number of batches helps in reducing costs of inventory management, storage and testing. There is a possibility of using this application in future for tomato, winter cauliflower, broccoli and cabbage as well. These seeds express diversified seed sizes and precision sorting can be a solution.

Table 6 : Effect of precision grading on hybridity, germination and recovery of usable transplants of eggplant hybrid seeds.

Bottom screen mesh size (mm)[†]	Hybridity (%)	Germination (%)	Usable transplants (%) [‡]
1.7 round	82.93	61	61
1.8 round	83.87	67	67
1.9 round	84.07	73	72
2.0 round	84.80	79	76
2.1 round	85.07	65	65
0.8 slotted	91.57	67	63
0.9 slotted	95.57	72	70
1.0 slotted	86.33	68	67
1.1 slotted	84.77	60	60
1.2 slotted	83.23	61	61
Control (no seed sorting)	81.00	61	59
Mean	**85.76**	**67**	**65**
SE	**0.46**	**1.35**	**1.45**
CD (0.05)[§]	**1.35**	**3.97**	**4.28**
CV (%)	**0.92**	**3.49**	**3.84**

[†] Top and middle screen mesh sizes did not vary among treatments and were 4.0 (round) and 3.5 (round) mm, respectively.

[‡] Percentage of healthy, well grown, saleable seedlings or transplants obtained per gram of seed after 10-14 d of sowing.

[§] Critical difference value (p d" 0.05).

5.2 Processing to improve seed germination or vigour

In any processing plant re-milling or re-processing can be done to improve germination or vigour. The procedure to follow depends on the germination before upgrading or re-processing. To improve germination we have re-process the lots by removing smaller sized seeds. The thumb rule followed here is -

    Germination between 85 - 89%   - remove 10% light weighted seeds.
    Germination between 80 - 84%   - remove 15% light weighted seeds.
    Germination between 75 - 79%   - remove 20% light weighted seeds.
    Germination between 70 - 74%   - remove 25% light weighted seeds.
    Germination   < 70%   - remove 30% light weighted seeds

In case of onion for example-the processing for improving germination is done as follows:

1. Sizing: increasing 0.20 mm round. The screen-sizes are between 1.80-2.80 mm round.

2. Gravity separator: The percentage of rejects per size depends on the current germ, and on the targeted upgrade germ expected. Also, when before milling germ is a bit more than 85%, perform the following steps.

    a. Screen-size: 2.00-2.20 : Take size fraction and remove $\pm$ 30% of the lightest seed on the gravity separator, then regravity the good seed into 4 fractions.

    b. Screen-size: 2.20-2.40 : Gravity separate into 4 fractions.

    c. Screen-size: 2.40-2.60: Germ < 85% (of the original batch) Gravity separate into 4 fractions. Germ > 85% is ok.

    d. Screen-size: 2.60-2.80: Germ < 83% (of the original batch): remove 15% of the lightest seeds on the gravity separator and make a separate batch. Also create a batch of the rest (fraction 2-5). Germination > 83% is acceptable.

5.3 Liquid density separation

This method of seed separation is very simple and widely followed in any processing plant. LDS can be done manually or by using a mechanical liquid density separator. The principle involved here that, seeds with densities less than 1.0 (that of water) would float and those with greater densities would sink. It is possible to improve the seed quality when other mechanical methods have failed. This method easy to adopt at farmers' level since the media used are not expensive and easy to prepare. Certain easily available materials like common salt, sugar, vinegar etc. can be further used to increase the density of water, thus increased scope for enhancing the seed quality.

5.3.1 LDS by manual method

Before LDS with different liquid-it is necessary to quantify density of the diff. Liquids. *How?*

Find the Mass of an empty Measuring Cylinder. Add a certain Volume of Liquid. Find the Mass of the Measuring Cylinder and Liquid. Calculate the Mass of Liquid.

Mass of Liquid = Mass of Measuring Cylinder and Liquid – Mass of empty Measuring Cylinder

Calculate Density of Liquid (Unit- g/cm3 or kg/m^3

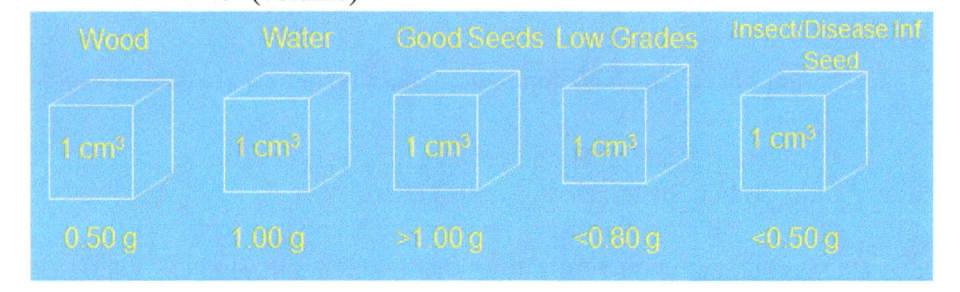

ρ (rho) = $\dfrac{m \ (\text{mass})}{V \ (\text{volume})}$

Wood	Water	Good Seeds	Low Grades	Insect/Disease Inf Seed
1 cm³	1 cm³	1 cm³	1 cm³	1 cm³
0.50 g	1.00 g	>1.00 g	<0.80 g	<0.50 g

Fig. 39 : Specific densities of various material

Fig. 40 : Liquid density separation in hot pepper

5.3.2 LDS by mechanical method

The seed is manually fed into the seed hopper, from where the seed is vibrated into the mixing funnel into which water (or another liquid) is added. The seed hopper is available in different configurations. From the mixing funnel the seed and liquid mixture is injected near the bottom of the liquid tank, thereby eliminating surface tension effects. An increased water pressure at the outlet improves the separation effect.

The process is based on differences in buoyancy of good, mature seeds and defective seeds. The good fraction of the seed lot as well as empty seed with complete and undamaged seed coats will float to the surface of the tank. The empty seed needs to be separated in the gravity separation process. When reaching the surface, the seed is transferred continuously by the tank overflow stream to the seed distribution tube and further into a seed collection box. The principle of the liquid separation can also be reversed to separate heavy seeds from lighter impurities. In this case the process is batch wise. The good seed is collected at the bottom of the tank, to be drained out in intervals. The seed dryer box is placed in a dewatering station, which allows excessive water to drain off the seed. The drained liquid is collected and returned to the reservoir tank. From the tank an adjustable flow of liquid is pumped back to the mixing funnel. The separation tank has a removable stainless screen insert to enable easy emptying of debris that sinks to the bottom of the tank.

Fig. 41 : Liquid density separation by mechanical method

REFERENCES

Agrawal, R.L. 1996. Seed technology. Oxford and IBH Publishing Company, New Delhi.

Hanumaiah, L. and H. Andrews. 1973. Effect of seed size in cabbage and turnip on the performance of seeds, seedlings and plants. Proc. Assoc. Offic. Seed Anal. 63:173-185.

ISTA.2011. International rules for seed testing. Int. Seed Test. Assoc., Bassersdorf, Switzerland.

Kalavathi, D. and K. Vanangamudi. 1990. Seed size, seedling vigour and storability in cluster beans. Madr. Agric. J. 77:39-40.

Komba, C.G., B.J. Brunton and J.G. Hampton. 2007. Effect of seed size within seed lots on seed quality in kale. Seed Sci. Technol. 35:244-248.

Pandita, V.K. and K.S. Randhawa. 1995. Influence of seed size grading on seed quality of some tomato cultivars. Seed Res. 23(1): 31-33.

Panse, U.G. and P.V. Sukhatme. 1985. Statistical methods for agricultural workers. 2nd ed. ICAR, New Delhi, India.

Srimathi, P. and Vanangamudi, K. 1993. Seed size grading in cowpea *Vigna sinensis* (L.), SAVI. Legume Res. 16 (1-2):144-146.

Chapter - 6

Seed Blending

The term seed blending is nothing but seed mixing done based on certain principles. It is simply combining two or more seed lots of same variety or hybrid of specified quality. The blending may involve lots which may have different quality in terms of germination. Usually individual seed lots are much better compared to blended lots but from the point of efficient inventory management blending is advisable. The main purpose of blending is to pool all the variable lots in to a larger bulk and mix the constituents effectively so that each pocket of seed has a uniform quality.

6.1 Why seed blending

1. It will improve the quality of the seed lots-by mixing two lots having minimum standards one can enhance the seed quality of the bigger lot.
2. Helps in better inventory management by reducing the number of lots- in case of tomato the average size of seed lots is 10-20 kg which may difficult for inventory management and quality control. Instead of having 10 lots of this size, by blending we can have larger seed lot which is very useful in inventory management activities like packaging and movement
3. Blending help in storage by reducing the space and location required. It is easy to store one big lot than ten.
4. Blending help in after sales service by tracking of seed quality in the field. One bigger lot is better for tracking and packaging than ten smaller lots.
5. Blending help in easy transportation from one processing plant and another if the target market segments are different. If seed lots of a variety were grown in south India but to be marketed in the north, it is better to have one bigger lot than ten which is easy for transportation.

6.2 Methods in seed blending

For good seed blending there are three methods and details of which given as below:

1. **Mechanical seed blending using a blender**: In this method seed blending is done using a mechanical blender or mixer. Here to achieve uniformity all the constituent seed lots will be mixed in a blender. The mixing involves movement of particle s induced mechanically or by gravity. The two seed lots are fed in to the feeder bins simultaneously and circulated which ensure uniform mixing of these lots. The constituent seed lots held in large storage bins are measured electronically and introduced in to the blender at required proportions. The blended seed then automatically packed as bulk which will be used for packing smaller pockets or tins later. A typical mechanical blender operation is given in Fig.42.

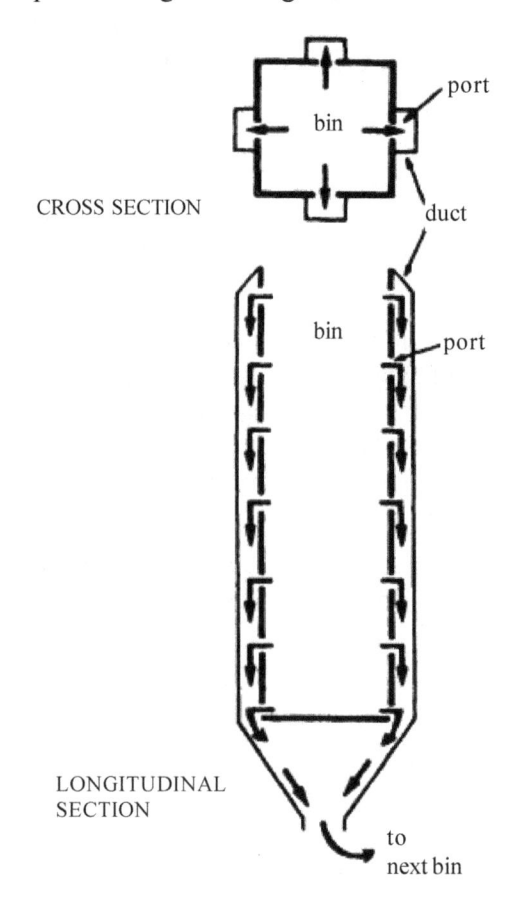

Fig. 42 : Operation of a mechanical seed blender

2. **Manual blending based on PLS (Pure Live Seed)**: The seed blending in the absence of large mechanical blender can be done following certain principles (detailed considerations given at the end of this chapter). One of the approach is based on pure live seed or PLS. This type of blending is useful when two lots have a similar quality and mostly similar quantity (+ or – <5%). Here is an example for this method:

Step-1: Calculation of PLS-Pure live seed can be calculated using the following formula:

$$PLS = \frac{Purity \times Germination}{100}$$

Step-2: Blending of lots : If the minimum PLS is 85%

Lot A: Seed lot having 98 % purity and 90 % germination gives 88.2 PLS

Lot B: Seed lot having 98% purity and 88 %germination gives 86.2 PLS

Then the blending of these two lots gives acceptable PLS

Seed Blend AB= (88.2+86.2)/2=87.2 PLS (accepted for blending)

3. **Manual blending based on germination %**: In this case the germination is considered when the seed lots to blend have different quality and quantity. In principle only two lots of having germination difference of more than 10% are accepted for blending. This approach is widely followed and useful for improving the quality of smaller lots.

Step-1: let us take minimum standard required is 90% of germination

Lot A: Seed lot having 50 kg weight and 95% germination

Lot B: Seed lot having 25 kg weight and 80% germination

Step-2: Blending of the lots

Lot A: 50kg X 95% of germination=4750.......1

Lot B: 25kg X 80% germination=2000...........2

Then add 1+2=we get 6750 and divided by total quantity i.e 75 kg we get 90 which is acceptable for blending since it is the standard fixed for the blend. We have improved seed lot quality of lot B from 80% to 90% by blending with bigger lot A having better germination quality.

6.3 Principles in seed blending

There are certain principles or considerations in seed blending which need to be followed to ensure the uniform quality in blended lot. Certain considerations are as below:

a) The seed lots should have similar PLS when blending of similar quality is done

b) The blender should be thoroughly cleaned and a dry run must be done to ensure free from other crop or variety seed.

c) The blending is done only for lots having declared results and meeting minimum standards like minimum germination.

d) For improving quality a minimum difference of 10 % should be considered between seed lots to be blended.

e) Before blending the colour and physical appearance (size, shape, texture etc) should be verified and matching quality lots only to be blended (Fig.43).

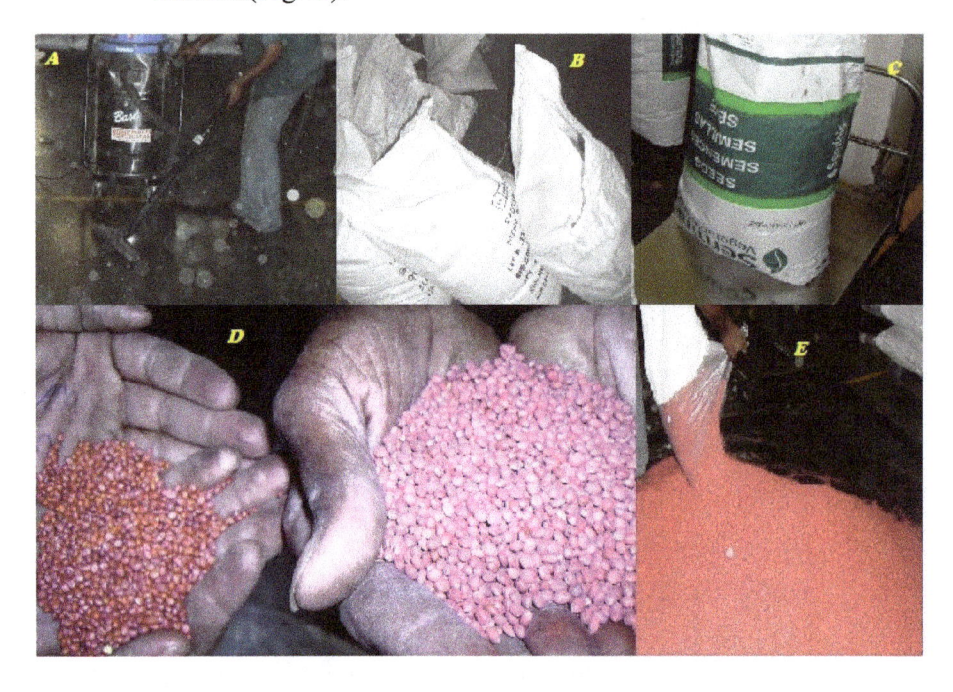

Fig. 43 : Steps in blending. A-Cleaning of the blending yard with vacuum cleaner; B & C-Getting all the bags and weighing of the seed lots to be blended; D-checking for the physical characteristics; E-Mixing of the seed lots

f) The international rules for seed testing prescribe a maximum size for

Seed Blending

blended seed lot as -10 tonnes (smaller seed) and 20 tonnes (bigger seed). In most of the cased size of the blended lot is decided based on quality and processing plant facilities.

g) Only authorised seed lots should be used for blending

h) A blended seed lot number should be created in inventory management system and to be followed during storage to packaging (Fig. 44).

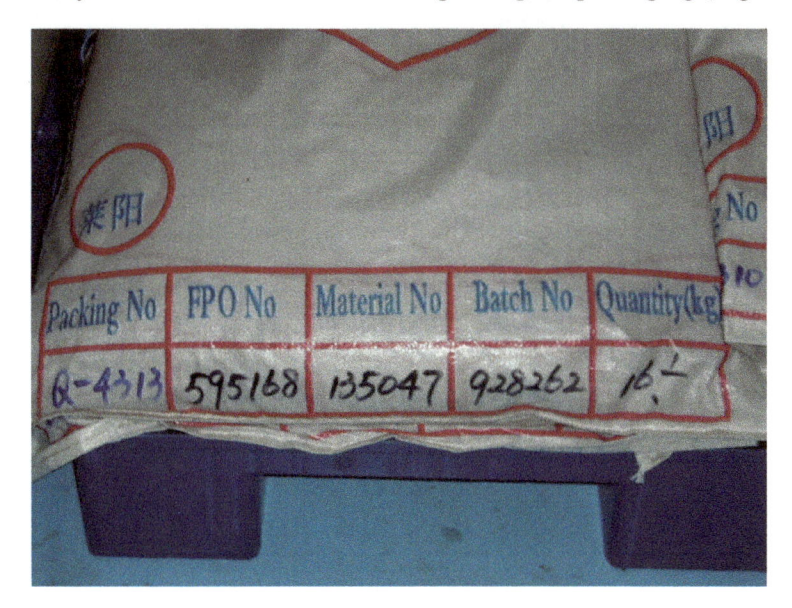

Fig. 44 : Bulk seed labelling for ease in lot tracking

i) Should blend two lots only at a time for easy tracking and lot numbers creation.

j) Re-processing, re-sampling or re-testing can be done on the blended lot for ensuring the quality before packing.

k) Constituent seed lots must be free from noxious weeds or seed borne pathogen. Seed health results are also considered before blending.

l) It is practice that blended lot is labelled as B-QUALITY but not always the case-it can be labelled as higher quality if all constituent seed lots qualify for the same.

CHAPTER - 7

Seed Treatment

Seed treatment is one of the important activities of the shop floor. The seed should be treated with various insecticides, fungicides and additives to protect it up to establishment. Some times enhance quality parameters like vigour which is critical for better establishment. Since vegetable seeds are of very high value and low volume, investment on seed treatment will yield better not only for the seed seller but to the grower also. More number of plants in a field brings more income per unit area. Hence the farmer or grower who buys seed is ready to bear the cost of seed treatment making it an important aspect of seed processing.

Seed treatment refers to the application of fungicide, insecticide or both to the seeds to disinfect(deep seated) and disinfest (over seed coat) them from seed borne or soil borne pathogenic organisms and storage insects. It also refers treating seed for enhancing quality by subjecting the seed to solar energy exposure or immersion in hot water or dry heat treatment or priming also.

7.1 Benefits of seed treatment

1. Prevention of spread of plant diseases both systemic and non-systemic. Seed treatments is effective in controlling bacterial, fungal and viral diseases. E.g-Treating tomato seeds on wet bases with Sodium hypochlorite gives protection against bacterial and fungal diseases.

2. Non systemic diseases that infect seed during harvest or storage period. E.g-Treating seeds with trisodium phosphate washes away the viral particles on the seed coat and hence give protection

3. Seed treatment protects seed from seed rot and seedling blights. Once the seed is planted the protective coating around the seed, acts as a barrier against seed borne and soil borne organisms.

4. Seed treatment improves the germination and vigour E.g- Priming improves germination and seed treatment with certain micronutrients improves seedling vigour.

5. Provides protection from storage insects and pests. For complete protection it is necessary to treat the seeds with insecticide also. E.g- In case of Sweet Corn treating seeds with metalaxyl or Vitavax
6. Seed treatment gives unique identity for the seed lot. Treating different seeds with different colour polymers can give better colour to the seeds. E.g-Treating seeds of gourds with golden colour not only cover the dull colour of these seeds but help in good marketing.
7. Helps in genetic integrity by better segregation in the storage: If two varieties or hybrids of same crop or different crops to be packed then we can use different colour polymers to distinguish them. This help in avoiding mechanical contamination. E.g-In case of cabbage or cauliflower
8. Helps in improving compatibility between different types chemicals and compounds. E.g-By using a neutral polymer we can treat chemicals like Bavistin and Potash Solubilising Bacteria together as different layers.

7.2 Types of seed treatment

1. **Seed dis-infection**: It refers to eradication of fungal spores present within the seed coat or more deep seated tissues. For effective control the fungicide must penetrate into the seed to kill the fungus.
2. **Seed dis-infestations:** It refers to the destruction of surface borne organisms that contaminated the seed surface but not infected the seed. Chemical dips, soaks, sanitation, fungicides applied as dust, slurry or liquids have been found successful.
3. **Seed protection**: To protect the seed and young seedling from organisms in the soil which might otherwise cause delay of the seed before germination.
4. **Seed Enhancement:** Seed treatment to improve lustre, quality and performance like polymer coating or priming

7.3 Scope for seed treatment- Must Do's

1. **Injured seeds**: Seeds suffer mechanical injury during threshing, drying or processing. Any break in the seed coat offers an excellent opportunity for the fungi to enter the seed and either kill it or weaken it. In this case a polymer film coating will be helpful which act as barrier for penetrating pathogen.

2. **Diseased seeds:** Seed may be infected by disease organisms at the time of harvest or during processing in storage. To prevent the pathogen to cause disease in later stages treating of suspected lots may be helpful.
3. **Undesirable soil conditions:** Seeds are sometimes planted under unfavourable soil conditions such as cold and damp soils, which favours the growth and development of certain spores enabling them to attack and damage the seeds.

Must Don'ts

1. **Fish mouth:** Seed treatment is not advised in some crops where seeds have split seed coat which are called as "fish mouth" as in case of certain pepper seeds. These seeds need to be packed naked and need to be treated with only recommended organic disinfectants.

7.4 Compounds used for seed treatment

A) **Mercurial Compounds:** like **Organo mercurials** (Phenyl Mercuric acetate (PMA) Methoxy ethyl Mercury chloride (MEMC) Ethyl Mercuric chloride (EMC) or **Inorganic mercurials** (mercuric chloride, mercurious chloride and mercuric oxide) are no longer be used since they are hazardous to people.

B) **Non Mercurials**-still in use but to be used with proper ESH (environment, safety and health) measures.

   1. **Organic Non Mercurials**: such as thiram and captan are widely used. They are less effective than the organic mercurials less damaging to the seeds and less dangerous to the persons handling the seeds. These fungicides act as seed disinfestants and or seed protectants. Over dosage is not harmful and viability is not affected. Eg: Thiram, Captan, Carbendazim, Metalaxyl.

   2. **Inorganic Non Mercurials**: Copper carbonate, Copper sulphate, Cuprous oxide are the major inorganic Non Mercuric compounds used as fungicides. Copper carbonate and Copper sulphate are used on wheat for prevention of bunt diseases. Cuprous oxide prevents seed decay and damping of in vegetables.

C) **Insecticides:** Though seed treatment pesticides are not used routinely but these can be used individually or in combination with different chemical and non-chemical compounds. Insecticides like imidachlorprid

(Gaucho), diazinon, chlorpyrifos, acephate, malathion, methyl parathion, and dimethoate.

D) **Polymers:** the use of polymers started since to overcome difficulty in uniform coating of the material on the seed when chemical slurry and to avoid loss of material which may be rubbed off the seed during transport and handling. In recent years, **film-coating** methods have been developed to overcome these problems. In film coating, the chemicals are applied in a polymer that is sprayed on the seeds as they fall though a specialized machine. The polymer is rapidly dried, so that the seeds emerge with a complete, dry polymeric coating. This coating does not rub off the seeds, greatly reducing worker exposure and waste. In addition, colors can be added to the polymers to code different varieties or seed lots. Large-scale units capable of treating large quantities of seeds have been developed, making it feasible to treat high-volume agronomic crops by this method as well.

Some novel applications have also been developed using the film coating method. For example, **artificial polymers** have been developed that exhibit temperature-sensitive permeability to water. These polymers are permeable to water at warm temperatures, but not at cool temperatures. Seeds coated with these polymers will not imbibe water – that is, absorb water and swell - if the temperature is below the set point of the polymer, potentially protecting the seed from chilling injury or germinating in an unfavorable environment. The coatings are also being used to delay germination after planting, such as for timing the emergence of parental lines at different times to ensure synchronous flowering for hybrid seed production. A **starch-based biopolymer** is also being used in film coating to slow water uptake and alleviate chilling injury.

Fig.45 : Various seed enhancement techniques

7.5 The following are the different combinations of polymer and chemicals for seed treatment.

Table 7 : Seed Treatment chemicals and combinations for different vegetable seeds

Species	Target Diseases	Polymer Colour	Method/ Chemical	Alternate Chemicals	Insecticide a.i	Dose ml/kg seed	Water Dosage -ml
Beet Root	Black rot	Red	Thiram	Captan	-	-	5
Asparagus	*Fusarium* Wilt	Red	Captan	Bleach soak	-	-	5
Bean	Phythium/ Rhizoctonia	Green	Thiram	Streptomycin/ Metalaxyl S	-	-	10
Broccoli		Purple	Thiram	Hot water Soak	-	-	5
Carrot	Bacterial blight	Green	Thiram	Hot water Soak	-	-	10
Cauliflower	Alternaria, Black leg,	red	Captan	Hot water Soak	-	-	5
Cabbage	Black rot, Seed rot, Damp. off	red	Thiram		-	-	5
Coriander	Damping off	red	Thiram	-	-	-	15
Eggplant	Fruit rot	red	Thiram	Hot water soak	-	-	5
Hot Pepper	Anthracnose	red	Thiram	Bleach soak	-	-	5
Khol Rabi	-	red	Thiram	-	-	-	10
Lettuce	-	green	Thiram	-	-	-	10
Okra	Damping off	red	**Metalaxyl S**	Thiram	Imidacloprid 48% FS	8.0	10
Onion	Damping off	green	Thiram	Methocel	-	-	5
Radish	Bacterial blight	puple	Thiram	-	-	-	5
Ridge Gourd		golden			-	-	15
Sponge Gourd		golden			-	-	15
Cucumber Squash Water Melon	Seed Rot, damping off, *fusarium* foot rot, black rot	red golden Red or golden	Captan	Thiram	Imidacloprid 48% FS - Imidacloprid 48% FS	5.0 - 10.0	15 15 15 15
Bitter Gourd		Golden	Thiram	-	-	-	15
Bottle Gourd		Golden	Thiram	-	-	-	15
Sweet Corn	Damping off	red	Captan D	Metalaxyl S	-	-	10
Sweet Pepper	Bacterial Spot	red	Thiram	Hot water/ TSP	-	-	5
Tomato	Bacterial spot, tobamo	red	Thiram	TSP/Mancozeb	-	-	5

7.6 Seed treatment methods

For vegetable seeds two important methods are followed in seed treatment.

Dust Treatment Method: In some of bulky vegetable seeds like coriander or onion dust method is followed. Melathion or Thiram are most widely dusts. Here place the seed and fungicide in drum seed treater and on mechanical agitation seeds will be uniformly coated with dust. For dust treatment the seed coat texture should be rough so that to ensure better coating. In a drum seed treater make sure that correct amount of fungicide is being applied to the seed flow or dump. Also it is important to ensure correct quantity of seed dump with measuring cup.

Slurry Treatment Method: For most of the vegetable seed slurry treatment is most preferred since treatment of polymer is also mandatory for these vegetables. Here the fungicide, water and polymer are mixed to make slurry. This slurry will be mixed manually (as in semi-automatic seed treater) or mechanically (as in automatic seed treater) with a measured seed dump. After stirring and swirling the seeds are thoroughly coated. Dry the seed before planting.

Fig. 46 : Polymer film coating and seed treatment (featured crop-coriander)

7.7 Equipment for seed treatment

1. **Semi Automatic Slurry treater**: This type of seed treater is used for both dust as well as slurry seed treatments. Here the calculation of fungicide, water, polymer and seed dump will be done manually. All these are added in a drum and with the help of measuring cup added to the seed dump of specified quantity. Only the stirring and releasing are automatic after standard time duration. Here the seed dump quantity and treating liquid measurement are very important. Used for very small batches of seed lots.

Seed Treatment 79

Fig. 47 : Manual seed treatment

2. **Automatic Slurry-cum-polymer treater**: In this case of seed treater everything is mechanical and automatic. From preparation of liquid to the seed dump. This machine has specific controls for each of operations. The treatment liquid is applied as slurry in accurately metered through mechanism containing slurry cup and seed dump pan. The cup introduces a given amount of slurry with each dump of seed in to mixing chamber where they are blended. The main features of this type treater:

a) The introduction of fixed amount of slurry to a given weight of seed

b) To obtain given seed dump weight these treaters are equipped with a seed gate that controls seed flow to the dump pan

c) The amount of treatment material applied is adjusted by slurry concentrations and the size of the slurry cup or bucket.

d) The mixing chamber is fitted with an auger type agitator that mixes the seed and moves it to the bagging end.

Fig. 48 : Automatic seed treater

7.8 Wet seed treatment

To ensure seed free of seed borne pathogen seed sanitation on wet bases is done when seed is received at the processing plant. The seed is treated with sodium hypochlorite and tri-sodium phosphate in different mixing chambers. The seed is treated with these solutions for a specified time duration and washed with clean water and sent for drying.

Fig. 49 : Wet seed treatment facility for Tomato

7.9 Seed quality enhancement techniques

Seed enhancements are the range of activities performed on the seed after its harvest and conditioning. These techniques will be performed to prepare seed for uniform sowing and to get uniform plant population. The seed enhancements can be performed on wide range of seeds which have distinguishable physical characters such as color, shape, texture etc. There are numerous methods of seed enhancements some of which were discussed here.

The objectives of seed enhancement are:

1. To improve germination and vigor of the seed.
2. To improve emergence and to ensure maximum plant stand establishment.
3. To induce stress and salt tolerance.
4. To facilitate precision planting or sowing.
5. To deliver seed protecting pesticides and nutrients.
6. To improve packaging efficiency by ensuring uniform physical characters.

There are old and new methods of seed treatments which were in practice all over the world such as:

Seed Treatment

1. **Salt Tolerance Technique**: Certain vegetables varieties grown in acidic and saline soil conditions need to be hardened for salt tolerance. The salt tolerance can be induced by seed treatment with certain chemicals like Menedione Sodium bisulphate which can induce better adaptation for these conditions.

2. **Seed Hardening Method**: Hardening, also called wetting, drying or hydration dehydration, refers to repeated soaking in water and drying at 15-25°C (Pen Aloza and Eira, 1993). Increased á-amylase activity and sugar contents were also reported in the hardened seeds (Basra et al., 2005). Seed hardening is most effective technique for vigor enhancement and establishment. Hardening of normal and naturally aged coarse rice seeds improved the germination and seedling vigor.

3. **Seed Encrusting**: Encrusting (aka 'minipelleting' or 'coating') applies less material, so the original seed shape is still (more or less) visible. When the added material is slight, the product may appear Similar to film coated seed.

Fig. 50 : Seed encrusting for irregular seed shapes

4. **Seed Film Coating**: Film coating of materials can be tailored to regulate seed water availability and gas exchange, thereby controlling timing of germ and emergence (and potentially avoid poor environments for crop establishment). In this method water-attracting materials can be incorporated, aids to imbibition and increased seed-soil contact. Temp sensitive film coatings can delay imbibition until a set temp is reached which will avoid soaking injury in large-seeded legumes, high-sugar sweet corn, also useful for better **nicking** in hybrid seed production. Oxygen-generating materials (e.g. Ca or Mg peroxide) can be added to supply more O2 in waterlogged environments. (Halmer, 2006)

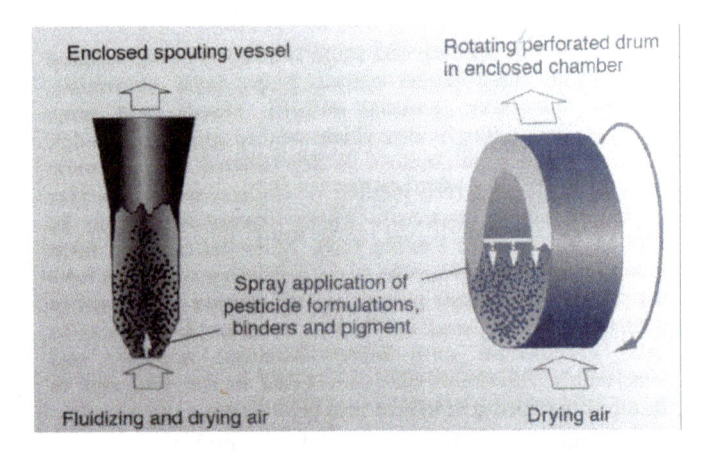

Fig. 51 : Seed film coating

5. **Seed Pelleting**: Enhancements used to (1) improve drill/seeder performance by making seeds rounder, (2) hange size grades, and (3) to carry nutrients, growth-stimulants and/or crop protection chemicals. Pelleting is typically used to round out small or irregular shaped seed, or to make small seeds larger – improves singulation and speed of sowing. Seed in rotating drum is wetted, and blends of powered materials (e.g. chalk, clays, perlite, lime, peat, talc) plus water-attracting or hydrophobic materials are progressively added, along with more water, until desired pellet wt or size increase is achieve. Wet-coated seed then dried with heated air, usually in separate equipment.(Halmer, 2006)

Fig.52 : Procedure in seed pelleting for irregular shaped seed

6. **Seed Priming**: A broad term in seed technology, describing methods of physiological enhancement of seed performance thru pre-sowing controlled-hydration methodologies. Seed priming also describes the biological processes that occur during these treatments. Primed seed contributes to better seedling establishment especially under sub-

optimal conditions at sowing (e.g. temperature extremes, excess moisture). Primed seed can also improve the percent useable seedlings in greenhouse production systems (e.g. plugs,(Bradford, 1984) transplants).

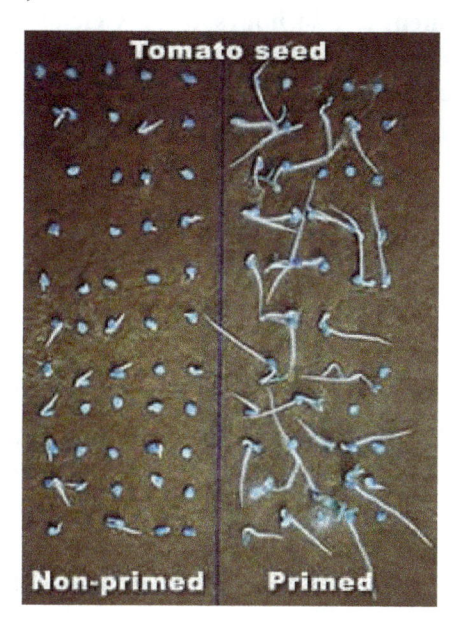

Fig. 53 : Effect of priming on tomato seed quality enhancement

7. **Seed Pre-germination**: Fully imbibed seeds germinated to point of visible radicles, then sorted and gradually dried to induce desiccation tolerance. Can produce damp pre-germ seeds (~30-55% SMC) with storage life of a few weeks at ambient temps, or dry pre-germ seed (viable for a few months). Commercially available at present for high-value flower seeds only (e.g. Impatiens, pansy). (Halmer, 2006)

Fig. 54 : Seed pre-germination technique for high value seeds

2. Seed Thermal Treatment: The seeds treated with hot water or dry heat treatment to disinfect from certain viral pathogen like CGMMV. This method can also be used along with chemical methods. The procedure to be followed as follows: Before the heat treatment the seeds must be dried back to **30% RLV. Length of the treatment 3 x 24 hours @ 76°Celcius** (this is **not** included the time that's needed to heat up the heat treatment cabinet). The temperature must be gradually increased from 30°C to 76°C:

The information below is just an example:Starting on Monday at 8.00 A.M.

 : Step 1 on 30°C - 30 minutes
 : Step 2 on 45°C - 90 minutes
 : Step 3 on 55°C - 90 minutes
 : Step 4 on 65°C - 90 minutes
 : Step 5 on 76°C - 60 minutes heating up + 72 hours

So on Thursday 2.00 P.M. the process can be stopped.

3. After the heat treatment the seeds must be stored for at least 4 days under ambient conditions in the conditioning Department (than the moisture level can get back to a "normal" level +/- 30% RH / 6.5% MOT)

4. After those 4 days we take the germ. samples, and the seeds are stored in the conditioned storage.

7.10 Precautions in seed treatment

1. Extreme care is required to ensure that the treated seed is never used for animal or human consumption.
2. Care must be taken to treat the seeds at correct dosage.
3. Do not treat the seed with high moisture content as it may be injured when treated with some of the concentrated liquid products.
4. Care must be taken use necessary safety accessories while seed treatment and follow ESH drill before start of any operation.

7.11 Causes of poor treatments

1. **Wrong fungicides.** Use of inappropriate fungicides, old dusts, etc., may prove relatively ineffective for protection against soil fungi.
2. **Inadequate dosages.** Failure to get sufficient fungicide on the seed results in poor seed treatment.

3. **Carelessness.** The use of the best available fungicides and the latest equipment for treating seeds does not by itself guarantee proper seed treatment. Adequate care is necessary regarding machine adjustments, etc., to treat seeds effectively.

7.12 Cost of seed treatment

The details of cost of treatment for less expensive seed using polymer is given below. Cost for all the crops can be worked out taking this as a reference crop.

Table 8 : Estimation of cost per treatment

Sl. no.	Particulars	Cost-Rs/kg	Remarks
1.	Polymer cost(recommended dosage-15ml/kg)	9.00	polymer cost per kg-Rs 600(price+excise+VAT)
2.	Cost of labour (per kg)	2.00	
3.	*Cost of packaging (per kg, pouch, labour, shipper)*		
	Pouch of 500 gm packing	7.00	
	Labour	2.00	
	Shipper	2.50	Rs 25 for 10 kg
4.	Cost of electricity	2.00	approximate
5.	Misc.costs	0.00	nil
	Total-Cost per kg	**24.50**	
Marketing List price (Rs per kg)		**148.00**	

Chapter - 8

Seed Drying

Vegetable seed drying is the most critical stage for any shop floor activities. Only properly dried seed will be taken for processing and all the other activities from seed receipt to storage. Seed producers harvest the seed which will have more moisture than packaged seed. For example onion seed harvested when its moisture at 60% but eventually dried to 10-12% for safe storage and packaging. Seed drying is highly skilled and specialized activity without affecting the seed quality. Proper drying procedures are vital for storage and field performance of seed lots. Drying prevents germination of the seed, the growth of bacteria and mould and it reduces the conditions for development of insects. The moisture content of a seed is influenced by the humidity of the air because they are very hygroscopic in nature. The more moisture in the air, the higher the moisture content of the product. If seed are harvested during warm, humid weather, the moisture content of the seed will be high because the relative humidity of the air is high as well. It is necessary to dry the seed before it can be stored safely. The method of drying depends on the local conditions (climate, season, volume of the seed, type of seed etc)

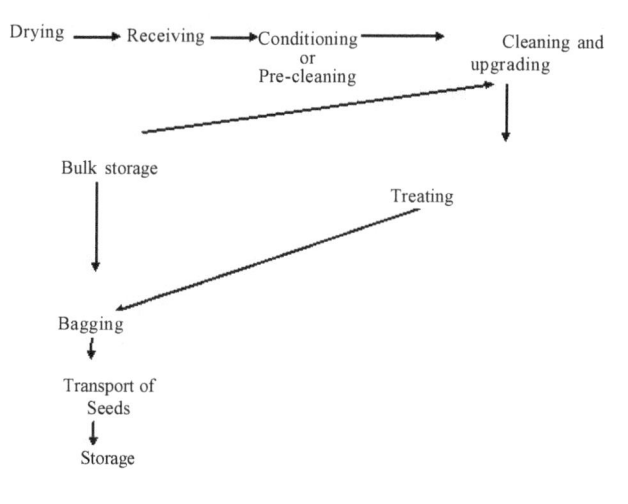

Drying ⟶ Receiving ⟶ Conditioning or Pre-cleaning ⟶ Cleaning and upgrading ⟶ Treating ⟶ Bulk storage ⟶ Bagging ↓ Transport of Seeds ↓ Storage

Fig. 55 : Operations from seed drying to storage

8.1 Principles of drying

Seed drying is a normal part of the seed maturation process. Some seeds must dry down to minimum moisture content before they can germinate. Low seed moisture content is a pre-requisite for long-term storage, and is the most important factor affecting longevity. Seeds lose viability and vigor during processing and storage mainly because of high seed moisture content (seed moisture greater than 18%).

High seed moisture causes a number of problems

1. Moisture increases the respiration rate of seeds, which in turn raises seed temperature. For example, in large-scale commercial seed storage, respiring seeds may generate enough heat to kill the seeds quickly, or to even start a fire if not dried sufficiently..
2. Mold growth will be encouraged by moisture, damaging the seeds either slowly or quickly, depending on the moisture content of the seeds. Some molds that don't grow well at room temperature may grow well at low temperatures causing damage to refrigerated seeds. In such a case there may be no visual sign of damage.
3. Unless seed moisture is at least eight percent or below, insects such as weevils can breed causing rapid destruction of seeds in a short period of time

Relationship between seed moisture content, temperature and humidity

Seed can be stored for long-term sealed storage provided that the seed moisture content is less than 8%, which means that the relative humidity must be kept below 35%

Seed Drying

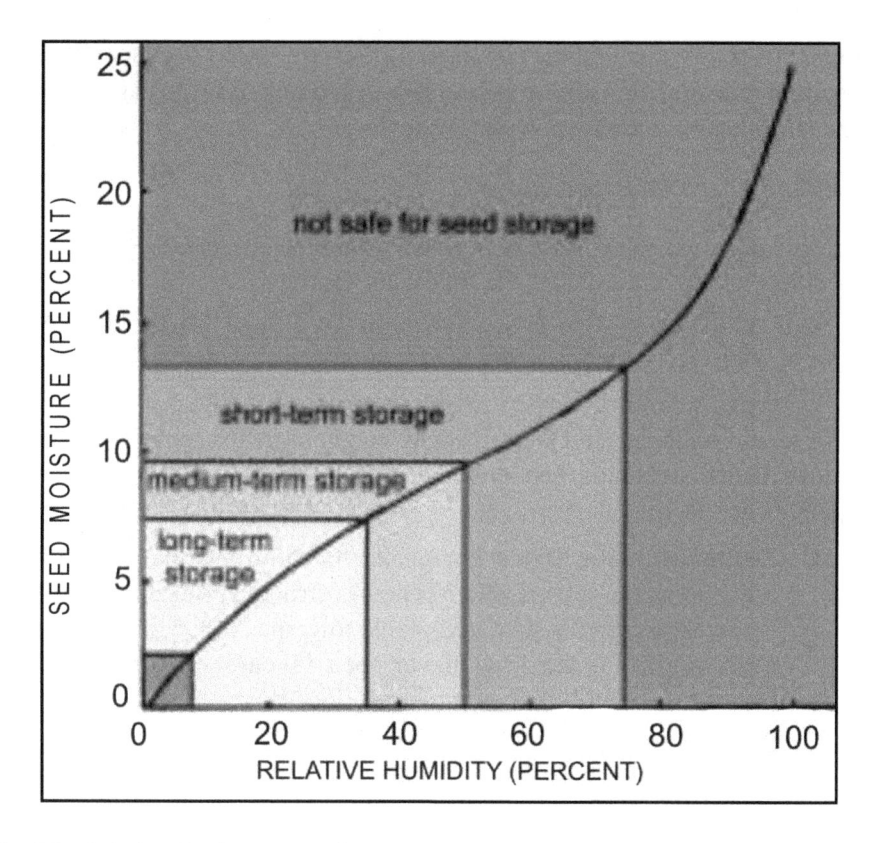

Fig. 56 : Relationship between moisture, temperature and relative humidity and effect on storage

The effects of temperature, moisture, and relative humidity were discussed above as separate factors which affect the longevity of stored seed. In reality, the effects of temperature and relative humidity are highly interdependent in their effect on stored seed. There is a simple method for calculating the combined effects of relative humidity and temperature on seed longevity, which is as follows: the sum of the storage temperature (in degrees F), plus the relative humidity (in percent) should not exceed 100. Since seed moisture is the most important concern, the rule stipulates that no more than half the sum should be contributed by the temperature (Harrington, 1960). The majority of crop seeds lose viability quickly when the humidity approaches 80% at temperatures of 77°F (25°C) to 86°F (30°C), but when stored at a relative humidity of 50% or less, and a temperature below 41°F (5°C), seeds will remain viable for at least ten years (Copeland, 1976). If seeds are taken from a cold or frozen storage and transferred to room temperate, care must be taken to prevent condensation on the seeds. If the seeds are in a sealed container, allow them to sit until they reach room temperature before opening the container. If they are stored in paper, place the seeds into a plastic bag with the excess air sucked out, seal the bag,, and wait for the temperature to stabilize before unsealing.

Determining seed moisture content

Moisture content of seed is defined by the International Seed Testing Association (ISTA) according to the following formula:

Seed Moisture = Fresh Seed Weight-Dry Seed Weight/Dry Seed Weight X 100

In order to determine the percent moisture content of fresh seed, a sample of fresh seed is weighed, and then an equal weight of fresh seed is dried slowly to remove the moisture, and then re-weighed. To quantify seed moisture various equipments are used.

a) **Digital Moisture Meter**: For quicker moisture determination a digital moisture meter is used which is non-destructive. But some time with some variations but with acceptable tolerance limits of +or – 0.5 %. This method is used whenever seed inventory is sent for final packaging. At the processing shop floor this will serve as second check. This equipment work on the principle of dielectric properties of seed with changes in moisture content. The properties like dielectric constant which is ration of permittivity of a substance to the permittivity of free space.

Fig. 57 : Digital moisture meter to analyse seed moisture based on di-electric constant

b) **Heating Vapour Method**: This method is destructive one and seed has to grounded before determining the moisture. The equipment is called as *Computrac 2000 XL*. This method is very accurate but destructive. This method is the faster than the oven method with average time for analysis is <10 minutes. The principle involved here is grounded seed sample heated to temperature up to 275°C and measures vapour out of moisture volatilization in quickest possible time (< 2-3 minutes). .

Fig. 58 : Analysis of seed moisture based on vapour

c) **High Constant Temperature Oven Method**: This method is used in the seed testing laboratory and a bit time consuming but very accurate. Here the seed in a container is weighed before and after drying. The temperature in the oven maintained at 130-133°C. The samples for 1-4 hours. E.g in case of Tomato. The difference between the weights is the moisture content of the seed lot which can be calculated using the following formula.

$$M = \frac{(M2-M3)}{(M2-M1)} \times 100$$

Where M=Moisture %; M1=Weight of container and its lid in gm; M2=Weight of seed, container and its lid before drying; M3= of seed, container and its lid after drying.

Low Constant Temperature Oven Method: This method is similar to above method with difference in temperature which is maintained at 103+/-2°C and dried for 17+/-1 hours. The humidity maintained at <70 %. After the drying the sample is placed in a desiccator and moisture is determined.

Fig. 59 : Optimum seed moisture permits for safer storage

The maximum permissible seed moisture allowed in some of the crops as follows:

Type of seed	Maximum % Seed Moisture	Type of seed	Maximum Seed Moisture %
Beans	7.0	Beet	7.5
Musk melon	6.0	Cabbage	5.0
Onion	6.5	Carrot	7.0
Cauliflower	5.0	Celery	7.0
Sweet corn	8.0	Coriander	8.0
Sqash/pumpkin/cucumber	6.0	All other gourds	6.5
Tomato	6.0	Egg plant	6.0
Water melon	6.5	Hot Pepper/capsicum	7.0

8.2 Methods of seed drying

There are different methods for seed drying which can used as per the type of material and available space:

a) **Open Sun Drying**: This type of seed drying is commonly followed when seed material is very bulky and drying yard is available. The seed must be dried in shade and early hours of the forenoon to avoid injury to the seed. This material is not advisable for smaller and expensive seed (brassicas, tomato or peppers) but can be followed for bulky seed (okra, coriander). This method is highly dependent on the weather and hence not advisable. The seed drying parameters like moisture removal %, monitoring drying periods, temperature and uniformity in drying are not controllable and there is chances of remoistening during humid conditions. Here the heat is transferred by convection from the surrounding air and by absorption of direct and diffuse radiation on the surface of the seed. The convected heat is partly conducted to the interior increasing the temperature of seed layer which is used for effecting migration of water vapour from interior to the surface.

(a)

(b)

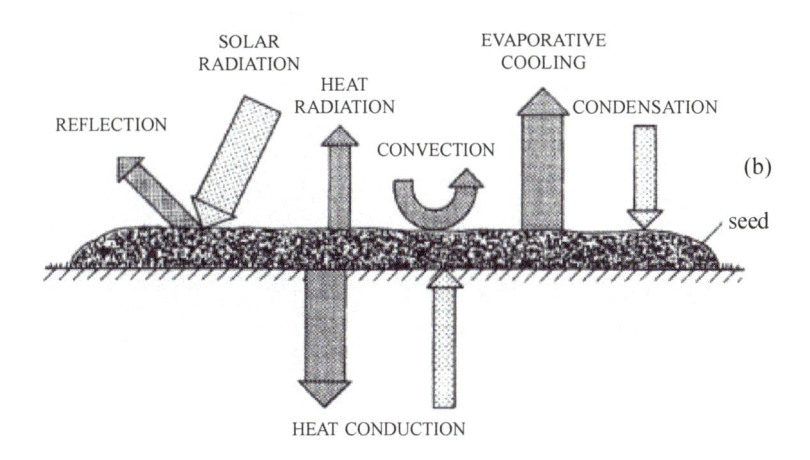

Fig. 60a & b : Open sun drying and principle involved

b) **Machine Drying**: This method is standard drying method in any processing plant. This is also known as forced air drying. Here there will be complete control on the temperature humidity and drying period. This method is used for safe drying of seeds to the desired levels of moisture for storage till packaging. Seeds cannot be dried by exposing them to heated air as elevated temperatures will destroy their germination potential. The alternative applied is to release the moisture from the product to the surrounding air. The dryer by maintaining the air at a lower moisture level can increase the drying rate. Moreover it removes the variability of weather as a factor in a drying operation. The dehumidifying seed dryer uses the following principle. The seed dryer essentially consists of two parts: the first is a chamber fitted with perforated trays vary with the quantity of seeds to be dried and also the size of the equipments. On top is housed the second part- the dehumidifying dryer, which continuously feeds dry air into the chambers. The removal of moisture is based on the principle of physical adsorption. It is found that to optimise drying capacity with minimum power consumption and to retain germination potential in the seed, the air coming out from the dry chamber should be at a temperature of 100°F+5%RH.

(a)

(b)

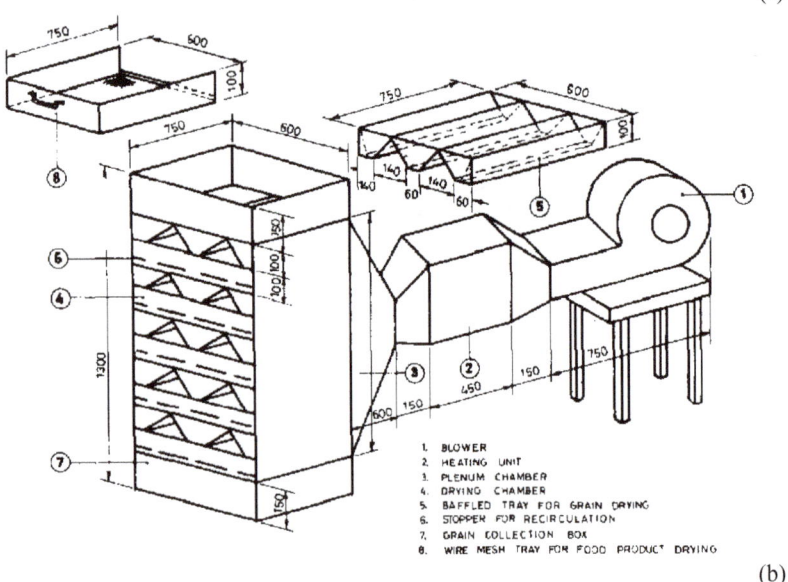

Fig. 61a & b : Operation of a mechanical seed dryer

c) **Drying using silica gel**: This type of drying is done for very small seed lots up to 10-100 kg. Very useful for high value seed which can be dried in several batches at once. Any hygroscopic substance that can be dried can act as a desiccant and absorb moisture from the surrounding air. If moist seeds are sealed in a container with a dried desiccant, the desiccant will dry the air, which will in turn dry the seeds. Silica gel is commonly used to dry seeds. Choose a non-porous container of appropriate size, with a tightly fitting lid which will seal effectively. Using silica gel to dry seeds Silica gel is available as clear bead or as indicating beads which change colour according to moisture status. Methyl violet indicator is dark green when wet and orange when dry. The colour change from wet to dry occurs either side of a 20-25% RH boundary. The principles developed by Royal Botanical Garden, Kew are given here.

- Dry the seeds to ambient conditions before you start.
- Fill the container approximately 20% by volume with oven-dried silica gel beads. A mix of 10% indicating to non-indicating beads is recommended.
- Put seed collections, held in cloth or paper bags, into the container, ensuring adequate air circulation.
- Maintain a minimum weight ratio of 1:1 silica gel to seed material.
- Place the drying container out of direct sunlight, in a cool place.

Fig. 62 : Seed drying using desiccants like silica gel

Types of drying from harvesting to packaging

Automatic Seed Dryer: Automatically drying is the basic installation in any processing unit with controlled RH and Temperature. The capacity of drying per hour is usually less in this case but useful for smaller seed lots

Box Drying: In this system the boxes are stacked in rows in front of the air-distribution system. The openings in the air-distribution system are aligned with the pallet openings at the bottom of the boxes. Air from the air-distribution system is blown through the boxes containing the product.

Individual Box Drying: With this box dryer every box containing seed is dried individually. Using this method the boxes can be placed in and removed from the dryer one by one. The drying of the seed can therefore start the moment the box is filled. The desired drying can be programmed for each individual box.

Silo drying for packaging seed: Drying online just before packaging with drying silo and vacuum transport. Dry seeds from storage will be loaded (by a separate vacuum system) into a mobile container or silo. The container as well as the silo is closed, so minimum moist will be absorbed.

Drying with dehumidified air: The centre of a dehumidifier is the rotor or absorption wheel. This part consist out of a chemical bound silica gel that can absorb moisture out of the air that pass the rotor but can also release moisture during the regeneration process.

Mist controlled drying: Drying seeds more economically to the desired moist content with your existing drying installation. If the average air temperature increase the absolute moisture content will also increase. Drying seeds to the equilibrium will not succeed when humid air is extracted. Dry air is needed to remove the last percentage of moist out of the seeds.

Chapter - 9

Seed Storage

Seeds are having a ability to survive and start a new generation when favourable conditions exists. However like other form of life, they cannot retain their viability indefinitely and eventually deteriorate and die. To retain viability and germination it is important to store in proper storage. Seeds harvested may not be useful for packaging thus making it to store for a period till the need for packaging arises. Seed storage is very critical task of seed processing. The success of seed processing is determined by safe storage. Since the vegetable seeds are expensive need modern techniques of storage. Based on the seed shelf life or longevity vegetable seeds can be divided into two types:

1. **Orthodox Seeds** : Orthodox seeds are long-lived seeds. They can be successfully dried to moisture contents as low as 5% without injury and are able to tolerate freezing temperatures. Most orthodox seeds come from annual temperate species adapted to open fields. At physiological maturity they contain moisture content of 30 – 50%. Most of the vegetable seeds are orthodox and hence there is lot of scope for proper storage. E.g-Tomato (6.0 %) and Peppers (7.0%)

2. **Recalcitrant Seeds** : They are short - lived seeds, which cannot be dried to moisture contents below 30% without injury and are unable to tolerate freezing. They are difficult to ore successfully because of their high moisture content encourages microbial contamination and results in more rapid seed deterioration. Storage of these seeds at subzero temperatures causes the formation of ice crystals, which disrupts cell membranes and causes freezing injury. These seeds are from perennial trees in the mois t tropics such as coconut, coffee, cacao, citrus etc. These seeds mature and exists in their fruits and are covered with fleshy or juicy ariloid layers and impermeable testa. At physiological maturity they contain more moisture content (50- 70%) than orthodox seeds, even though their embryos are only about 15 % of the size of an orthodox seed embryo. In general recalcitrant seeds never go into dormancy but instead continue their development and progress towards germination.

9.1 Factors affecting shelf life or longevity seeds

1. Genetic factors: Seeds of some species are genetically and chemically equipped for longer storability than other under comparable conditions. Most long-lived seeds belong to species possessing hard, impermeable seed coat. Generally seed species possessing high oil content do not store well as those with low oil content. Quantity of oil present in embryo portion of seed is responsible for storability. For eg. Lettuce can be stored for 6 years, muskmelon & radish for 5 years. Also beet, cabbage, egg plant and tomato seeds for 4 years.

Table 9: Approximate life expectancy of vegetable seeds stored under favourable conditions.

Vegetable	Years	Vegetable	Years
Asparagus	3	Kohlrabi	3
Bean	3	Leek	2
Beet	4	Lettuce	6
Broccoli	3	Muskmelon	5
Brussels sprouts	4	Mustard	4
Cabbage	4	New Zealand spinach	3
Carrot	3	Okra	2
Celeriac	3	Onion	1
Cauliflower	4	Parsley	1
Celery	3	Parsnip	1
Chard, Swiss	4	Pea	3
Chicory	4	Pepper	2
Chinese cabbage	3	Pumpkin	4
Collards	5	Radish	5
Corn, sweet	2	Rutabaga	4
Cucumber	5	Salsify	1
Eggplant	4	Spinach	3
Endive	5	Squash	4
Fennel	4	Tomato	4
Kale	4	Turnip	4
		Watermelon	4

Table modified from D. N. Maynard and G. J. Hochmuth, Knott's Handbook for Vegetable Growers, fourth edition (1997)

2. Initial seed Quality: The physical condition and physiological state of seeds greatly influence their life span. Seeds that have been broken, cracked deteriorate more rapidly than undamaged seeds. Several kinds of environmental stresses during seed development and prior to physiological maturity can reduce the longevity of seeds. For example deficiency of minerals (N,K,Ca), water and temperature extremes. Immature small seeds within a seed lot do not store as well as mature and large seeds within a seed lot. Hard seediness also extends seed longevity. When seeds not harvested at correct physiological maturity not store well even in favourable conditions. In case of tomato varieties pusa early dwarf harvested at Orange-1, PKM-1 at red-1 and Arka Abhijeet at red-2 have well developed embryo and endosperm. The seeds harvested at these stages can be stored for 2-3 years compared to other stages which lasted only 1-2 years. The delayed harvesting of seeds may result in mechanical injury during standard seed processing.

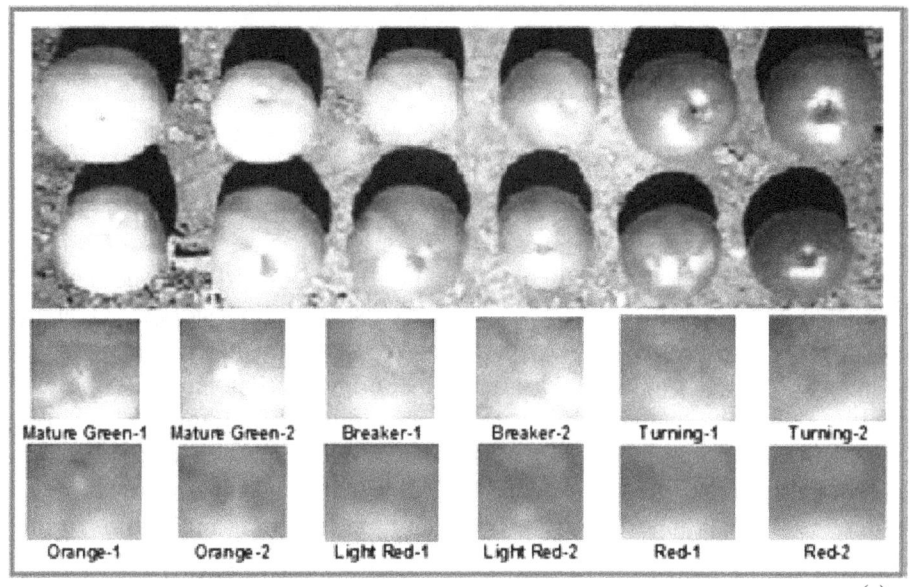

(a)

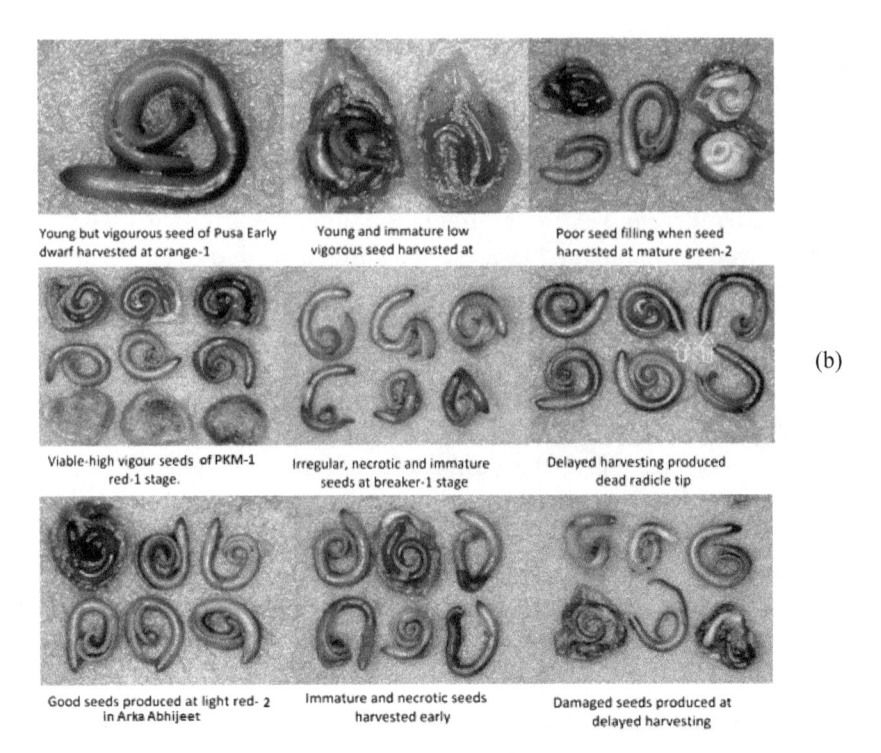

Fig. 63a & b : Effect of stage of harvesting on seed quality

3. Seed Moisture: Moisture content of the seed is one of the important factors influencing the viability of seed during storage. Over the moisture range, the rate of deterioration increases with increase in moisture. In general for every 1% decrease in moisture the store potential of the seed doubles (when the seed moisture is in the range of 4-14%). If the seed moisture content is in the range of 12- 14 %, losses occur due to increases mould growth and if the moisture content is above 18 -20% due to heating of the seed. Moreover within the normal range, biological activity of seeds, insects and moulds further increases as the temperature increases. The higher the moisture content of seeds the more they are adversely affected by both upper and lower ranges of temperature. At very low moisture content of 4 per cent seeds may be damaged due to extreme desicca tion, or breakdown of membrane structure hastens deterioration. This probably a consequence of reorientation of hydrophilic cells membranes due to loss of water molecules necessary to retain their configuration. Since the life span of seeds largely depends on the moisture content it is necessary to dry it to safe moisture limits before storage. However the safe moisture content again dependson length of storage, type of storage structure and kind of the seeds to be stored. For cereals in ordinary storage conditions for 12-18 months the seeds should be dried to 10 –12 % moisture content. However for storage in sealed containers (Hermetic packing) the seeds should be dried to 5 to 8 per cent moisture content.

4. Relative humidity and Temperature: the most important factors that influence the life span of seeds are relative humidity and temperature. The effects of R.H. and temperature of the storage environment are highly interdependent. Most crop seeds loose their viability at R.H. approaching 80% and temperatures of 25-30°C but can be kept for 10 years or longer at R.H. of 50% or less and a temperature of 5°C to lower (Toole 1950).

According to Harrington, 1973 because of interdependency the sum of the percentage of RH plus temperature in oF should not exceed 100 for safe storage. Harrington suggested the following thumb rules regarding optimum storage conditions.

1. For every 1% reduction in seed moisture the storage life of seed doubles
2. For every 10°F reduction in temperature doubles the life span of the seed.
3. The sum of relative humidity in percentage and temperature in OF° should not exceed 100.

The thumb rule applies to only when the seed moisture is in-between 4 and 14 %.

Calculation of storage effectiveness: Storage temperature must be derived from this formula when assessing storage unit effectiveness:

$$°F+RH < or = 100$$

Lower relative humidity means higher allowable storage temperatures, as long as their sum does not exceed a combined value of 100. For example, storage areas with an RH value of 60% need to be cooled to 40Ú F to enhance maximum storage potential. Higher temperatures can be especially damaging if the seed is stored at higher moisture contents. Seed stored with lower moisture contents may tolerate higher temperatures better. The relationship between seed moisture content and temperature (Thomsen and Stubsgaard, 1998).

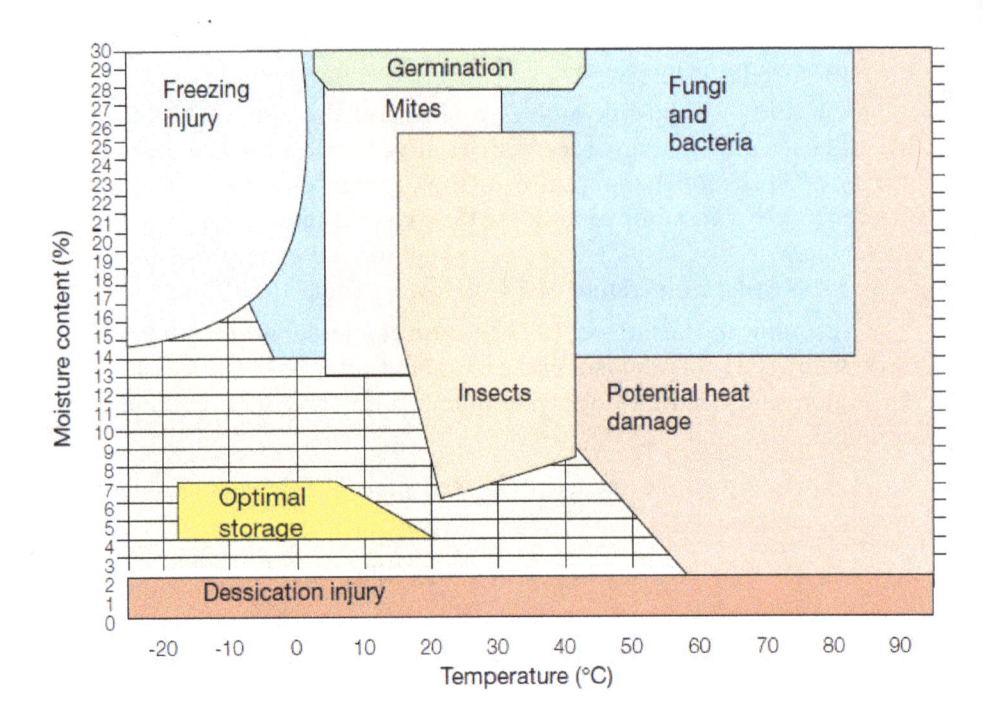

Fig. 64 : Seed storage affected by moisture content and temperature

5. Provenance: It has already been stated that a number of factors operating before and during harvest can affect seed viability. The samples obtained from different sources may show differences in viability behavior. The seeds harvested from regions of high relative humidity and temperature at the time of maturation or harvesting store less than the seed harvested from the regions of low relative humidity with moderate temperature.

6. Pre and post harvest conditions : Environmental variations during seed development usually has little effect on the viability of seeds, unless the ripening process is interrupted by premature harvesting, weathering of maturing seeds in the field, particularly in conditions of excess moisture or freezing temperature results in a product with inferior storage potential. Mechanical damage inflicted during harvesting can severely reduce the viability of some seeds eg. Certain large seeded legumes. Cereals are largely immune from mechanical injury presumably because of the protective lemma and palea. Small seeds tend to escape the injury during harvest and seeds that are spherical tend to suffer less damage than do elongated or irregularly shaped ones. During storage injured or deeply buried areas may serve as centers for infection and result in accelerated deterioration. Injuries close to vital parts of the embryonic axis or near the point of attachment of cotyledons to the axis usually bring about the most rapid losses of viability. High temperatures during drying or drying too quickly or excessively can dramatically reduce viability.

7. Oxygen Pressure during storage: Increase in oxygen pressure during storage tends to decrease the period of viability. Use of antioxidants has increased the storage period in some of the crops. If seeds are not maintained in hermetic storage at low moisture contents or even under conditions of constant temperatures and moisture the gaseous environment may change as a result of respiratory activity of the seeds and associated microflora.

8. Effect of storage conditions on the activity of organisms associated with seeds in storage

Bacteria : Bacteria probably do not play a significant role in seed deterioration. As germination is rarely reduced unless infection has progressed beyond the point of decay. Since bacterial populations require free water to grow, they cannot grow in stored seeds as the seeds are dry.

Fungi: Two types of fungi invade the seeds; field fungi and storage fungi. The field fungi invade seeds during their development on plants in the field or following harvesting while the plants are standing in the field.they cannot invade seeds during storage. Field fungi associated with wheat or barley in the field are *Alternaria, Fusarium, and Helminthosporium* spp. Storage fungi, mostly belong to the genera Aspergillus and penicillium. They infect seeds only under storage conditions and are never present before, even in seeds of plants left standing in the field after harvesting. Major deleterious effects of storage fungi are to decrease viability, cause discoloration, produce mycotoxins, cause excessive heat and develop mustiness and caking.

Insects and Mites: Deterioration of seeds by insects and mites is a serious problem, particularly in warm and humid climates. Weevils, flour beetles or borers are rarely active below 8% moisture content and 18-20°C, but are increasingly destructive as the moisture content rises to 15% and the temperature to 30–35°C. Mites do not thrive below 60% RH, although they have temperature tolerance that extents close to freezing. Hence for protecting the seeds from insects and mites the seeds should be stored at a moisture content of less than 10%, at a temperature of less than 20°C and the R.H. of less than 60%.

9. Other factors: Besides the above factors storage life is affected by number of times and kind of fumigation, effect of seed treatment etc.

9.2 General principles of seed storage

1. Seed storage conditions should be dry and cool
2. Effective control of storage pests

3. Proper sanitation in seed stores
4. Before placing seeds into storage they should be dried to safe moisture limits, appropriate for storage system.
5. Store only high quality seed i.e. seeds which are well cleaned, treated, with high germination and vigour.
6. Determine seed storage needs in view of period or length of storage time and preveling climate of the area during storage period.

9.3 Types of vegetable seed storage

There are various types of vegetable seed storage

a) **Open Dry Storage**: This type of storage is very short term from a week to 6 months. Generally this type of storages were used to store the seed immediately after receipt of the seed in the processing plant. There will no controlled conditions but equipped with ventilation and other standards like forklift path or rat proofing. The seed received if immediately to be processed then in this type of storage is very useful.

Fig. 65 : Uncontrolled open dry storage

b) **Cold Storage :** In this type vegetable seeds can be stored up to 12-15 months and mainly used for long term storage needs. These storages have temperature levels from 2°C to - 20°C with low humidity conditions. Once the seed has been dried to equilibrium with 15% relative humidity placed in a cold room where it is kept at a temperature of -20°C.

Fig. 66 : Controlled cold seed storage

c) **De-humidified (DH) storage**: In this type seeds are stored for 12-24 months. Here Moisture exclusion can be maintained in dehumidification systems. There are three types of dehumidification systems

Refrigeration type: This type of dehumidification work by drawing warm moist air on a metal coil through which a refrigerant (Freons like corbon-tetrachloride, trichloro-fluromethane etc.) is circulated. A part of atmospheric moisture is condenses on this cooling coil which is drained off. The cooled air coming from over the coil which now has a low temperature and a high relative humidity is reheated by the condenser coil of the refrigeration system; thus raising the temperature and lowering the relative humidity. The water removal capacity of this type of system is dependent on the difference in temperature between the entering air and the cooling coil. While these units are quite effective at high temperature, they lose efficiency below 70°F or 50% relative humidity. Heat from the electric motors that drive the compressor and fans add sensible heat to the atmosphere.

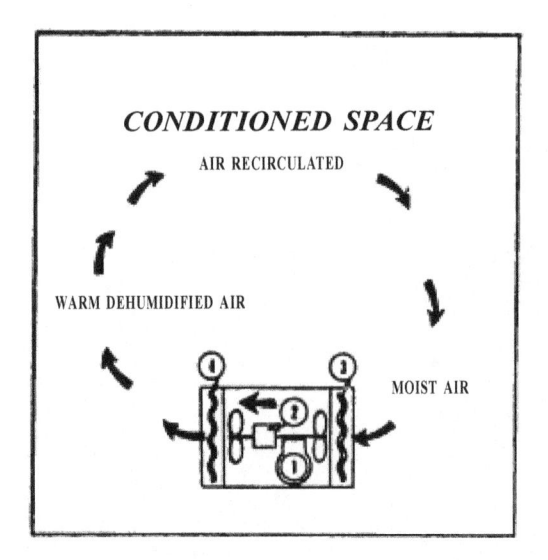

```
                    SYSTEM COMPONENTS
                    1. COMPRESSOR
  CONDITIONED SPACE  2. MOTOR AND FANS
                    3. EVAPORATOR COIL
                    4. CONDENSER COIL
```

DEHUMIDIFICATION SYSTEM, TYPE I

Fig. 67 : Refrigerated dehumidification seed storage

Chemical or Adsorption Type-The adsorption-type dehumidifier operates by drawing moist air over a solid drying agent (desiccant) which has the ability to extract and retain moisture on its surface by a phenomenon known as "adsorption." The air is filtered and dried to a very low dew point in the process, and the desiccant is periodically. Regenerated by means of heated outside air which vaporizes the moisture and dispels it to the outside of the conditioned space. Continuous operation of these machines is achieved by either using two desiccant beds which switch back and forth automatically, or by using rotating beds of desiccant, a portion of which is is always dehumidifying the air, while the remainder is being regenerated.

Desiccant dehumidifiers provide maximum efficiency at low temperatures, and are able to maintain constant relative humidity even below 10%. A factor that should not be overlooked is that heat is added to the controlled atmosphere even though the unit is placed outside the storage room. The latent heat of vaporization of the moisture that is removed is converted to sensible heat. There is also a certain amount of residual heat left in the desiccant after reactivation which increases the air temperature.

Seed Storage

CONDITIONED SPACE

AIR RECIRCULATED

MOIST AIR

WARM DEHUMIDIFIED AIR

DEHUMIDIFICATION SYSTEM, TYPE II

SYSTEM COMPONENTS
1. DESICCANT
2. HEATER COILS
3. BLOWER
4. REACTIVATION BLOWER

OUTSIDE AIR REACTIVATION

MOIST AIR

(a)

(b)

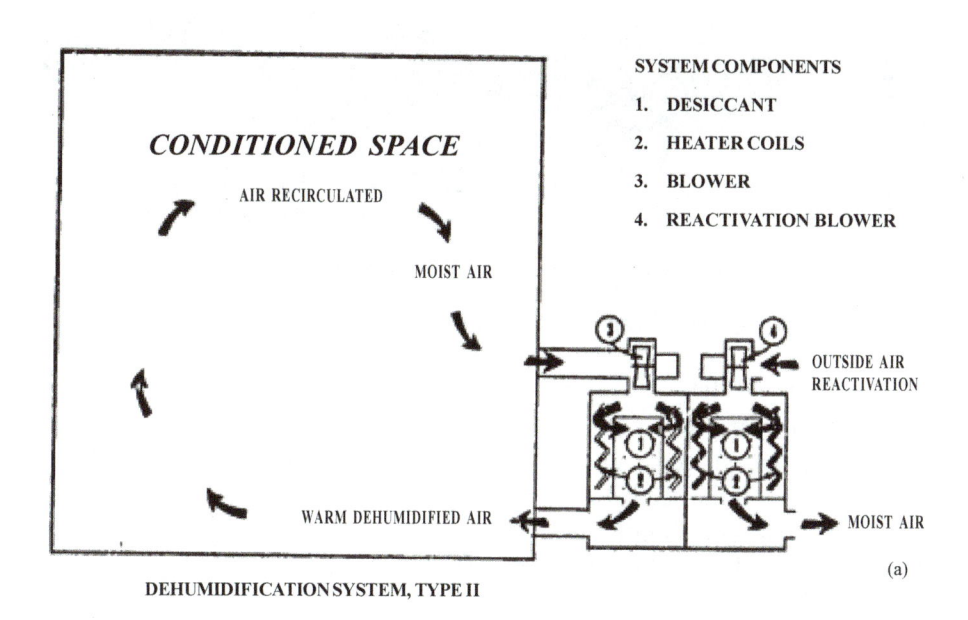

Fig. 68a & b : Adsorption type dehumidified seed storage

d) **Extreme Long term storage**: Here the small batches of vegetable seeds stored under extreme conditions like temperature using liquid nitrogen (-196°C) for a very long time up to 3-4 years. Also known as cryo storage. This method is used for stock seeds of parents or inbred lines used for breeding programme or also for short-lived seeds stored in liquid nitrogen vapour at approximately -196°C.

Fig. 69 : Cryo-seed storage under liquid nitrogen

9.4 Factors to be considered in storage

There certain factors to be considered while seed storage like:

a) Storing of seeds based on the volume and tracking the inventory: The bulky seeds and seeds of parental lines should be stored separately within a single storage. All the seeds lots stored should bear a stack card which help in tracking of tracking opening and closing balance. The stack cards give information whenever the seed lots used for different purposes like processing, treatment or packaging.

Fig. 70 : Inventory management in seed storage

b) The locations on the storage yard: Whenever the seed stored the locations should be labelled which help is tracking the seed lots better while choosing for different purposes. This helps in avoiding mechanical contamination or wrong packaging.

Seed Storage 109

Fig. 71 : Tracking of inventory in seed storage

C) Genetic integrity: The storage or processing yard should be free from cracks which can hold some seeds may result in mechanical contaminations. The crevises and cracks to be filled with white cement and periodically checked.

Fig. 72 : Genetic integrity or mechanical contamination issues in a processing yard

D) Fire extinguishers: The storages will hold expensive seed material for very long time. It is better to be equipped with approved extinguishers. A monitored fire detection system should be installed in all treating and storage areas, when and where available. Walls of storage area shall meet appropriate fire and building codes. In general A,B,E and F type extinguishers were used in seed storages. Also sand is standard extinguisher.

Fig. 73 : Fire safety accessories for a processing yard

CHAPTER - 10

Seed Inventory Management Systems

Inventory management system is a computer based system for creating work orders at a processing plant, tracking inventory levels in the go downs, , sales and seed movement out of the processing plant. These modern systems avoid book maintenance and save lots of space. It will also tool for avoiding overstock or delay in movement of seed material for packaging. In most of the processing plants SAP is the most popular inventory management tool which is more efficient than others. SAP is basically a tool of enterprise resource planning which help in material management. The customised software can be used very effectively in a processing plant for the following operations.

10.1 SAP Basic information & structure

1. **SAP R/3 system**
   - One of Famous ERP Package.
   - Can exchange all data per each function
   - Use Oracle Data Base.
     * ERP Package - Enterprise Resource Planning
2. **Attribute of SAP system.**
   - User can trace all changes in DATA processing.
   - Cannot change result simply, to change result

   Reverse all process sequence from end.

   c.f – Excel, Many companies own Database or Management program

10.2 SAP structure

This software is very versatile and can used from head quarter up to the sales offices. All the processing units can be linked to a central operations unit (fig. 74).

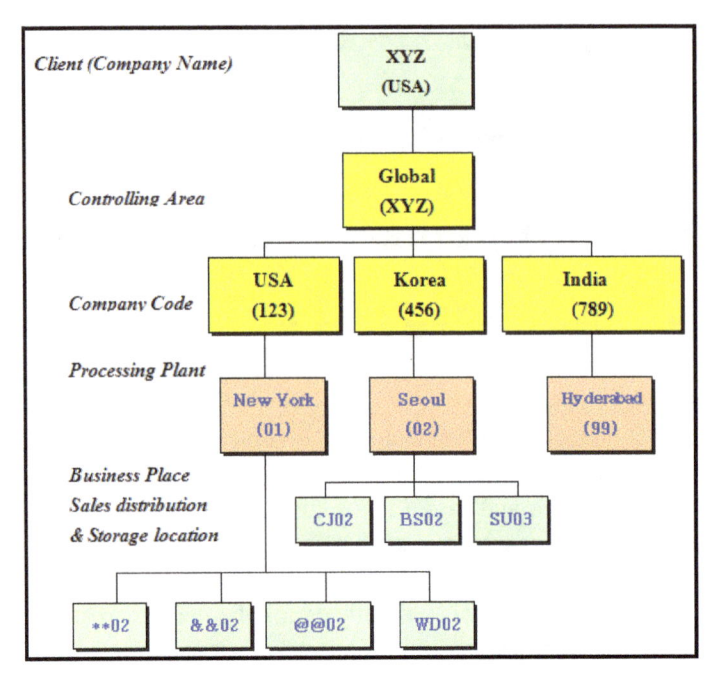

Fig. 74 : SAP inventory management structure

10.3 SAP functional modules

The SAP is having the functional modules in operations and seed processing (Fig. 75 & 76)

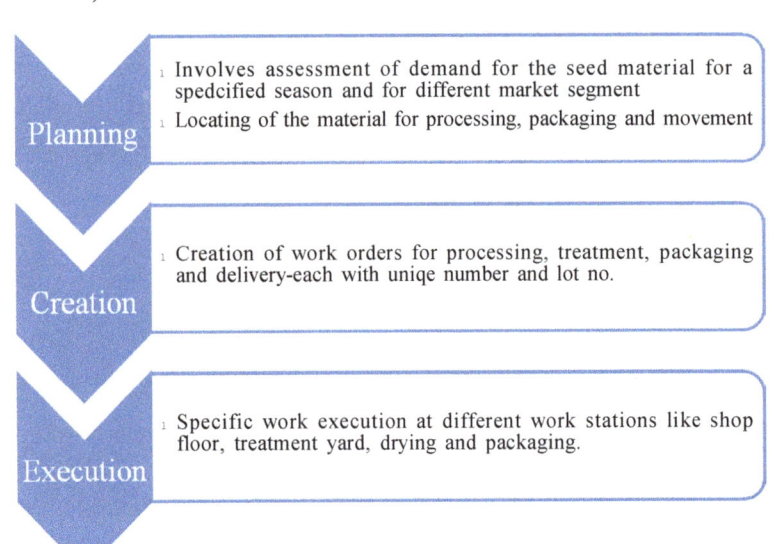

Planning
- Involves assessment of demand for the seed material for a spedcified season and for different market segment
- Locating of the material for processing, packaging and movement

Creation
- Creation of work orders for processing, treatment, packaging and delivery-each with uniqe number and lot no.

Execution
- Specific work execution at different work stations like shop floor, treatment yard, drying and packaging.

Fig. 75 : SAP inventory management functional modules

Seed Inventory Management Systems

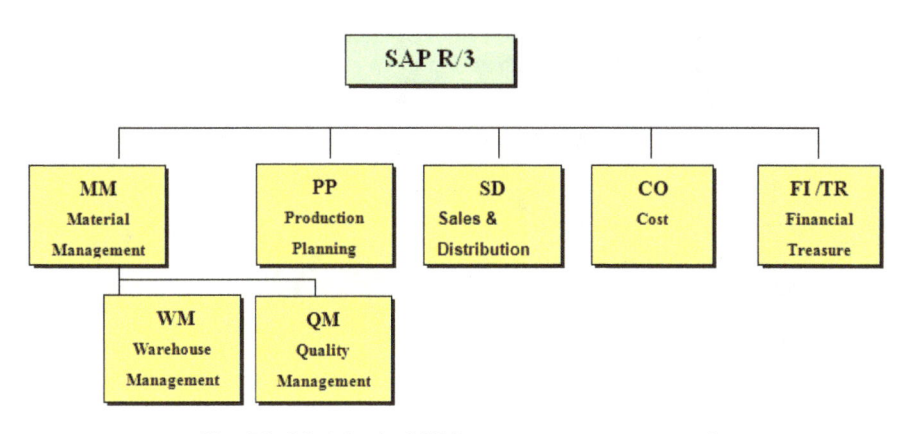

Fig. 76 : Modules in SAP inventory management tool

10.4 Screen structural elements

A typical SAP system in a processing unit work as centralized system of monitoring, processing orders, details of seed quality (batch inquiry) or up to delivery. An example of seed processing order created in SAP and its execution given below.

Function: Create processing or seed cleaning order which is called FPO (Field Production Order).

Step-1: The SAP screen structural elements: The SAP-ERP tool can be used for the various aspects of a processing unit. Each module have different functions. For detailed SAP applications and licensing contact the local office at Bangalore (call toll free-18002662208 or 080-66655771)

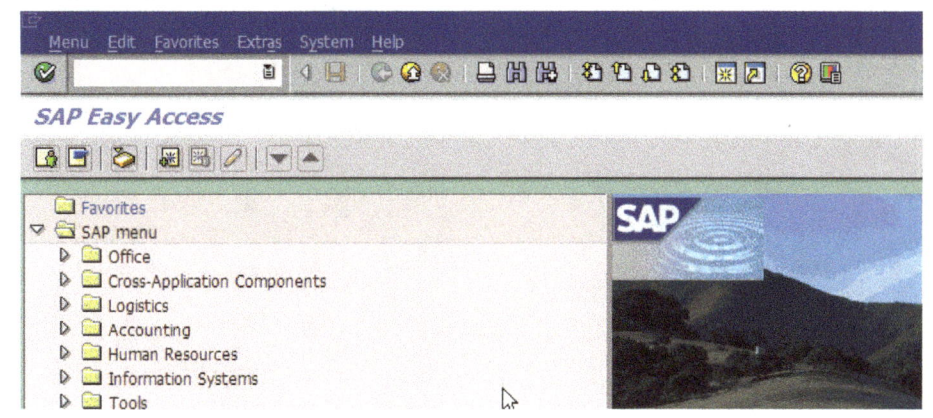

Step-2: Create an order to initiate and authorize processing: To initiate processing operation first locate the material number (130575) and create type of the order-here it is cleaning order (CD01). Also locate the material (SK01-processing plant in korea)

Production Order Create: Initial Screen

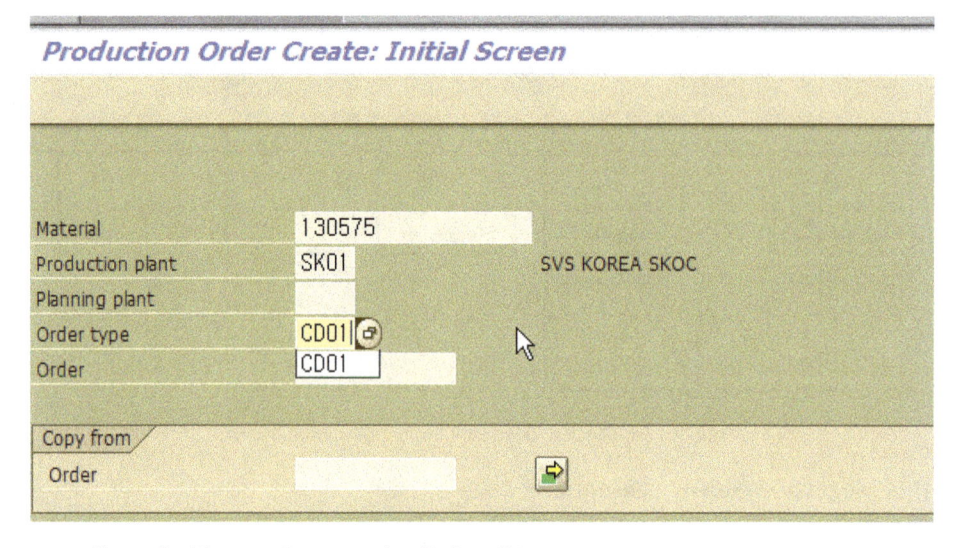

Material	130575	
Production plant	SK01	SVS KOREA SKOC
Planning plant		
Order type	CD01	
Order	CD01	

Copy from
Order

Step-3: Enter the required details-

- Enter material number, plant, order type
- Order type: "CD01" for cleaning, " PP01" for other operations.
- Enter details like quantity (e.g-77 kg)

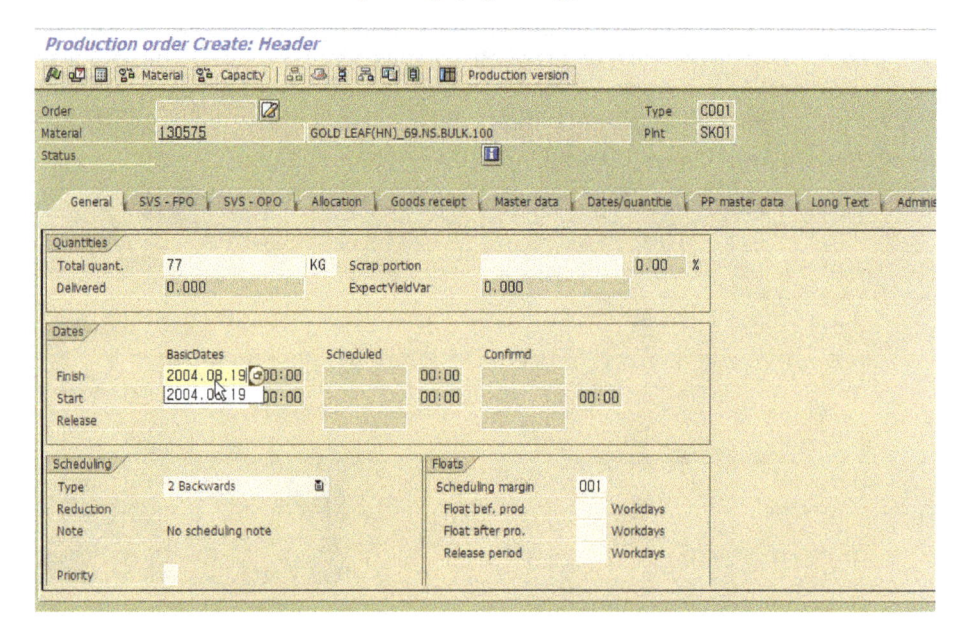

Production order Create: Header

Order Type CD01
Material 130575 GOLD LEAF(HN)_69.NS.BULK.100 Plnt SK01
Status

Quantities
Total quant. 77 KG Scrap portion 0.00 %
Delivered 0.000 ExpectYieldVar 0.000

Dates
 BasicDates Scheduled Confrmd
Finish 2004.08.19 00:00 00:00
Start 2004.08.19 00:00 00:00 00:00
Release

Scheduling **Floats**
Type 2 Backwards Scheduling margin 001
Reduction Float bef. prod. Workdays
Note No scheduling note Float after pro. Workdays
   Release period Workdays
Priority

- Enter processing type –in this case it is cleaning.

Seed Inventory Management Systems

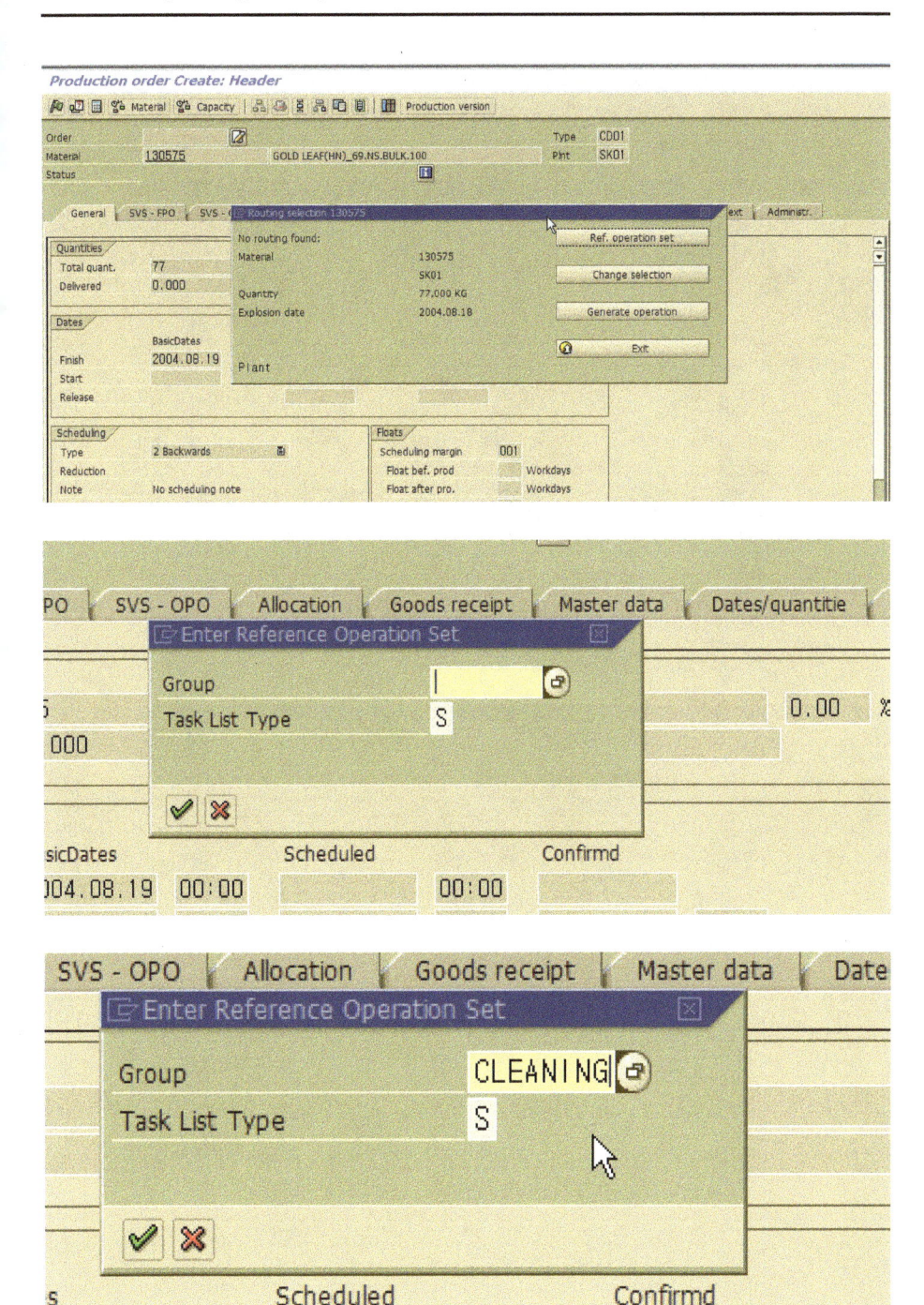

115

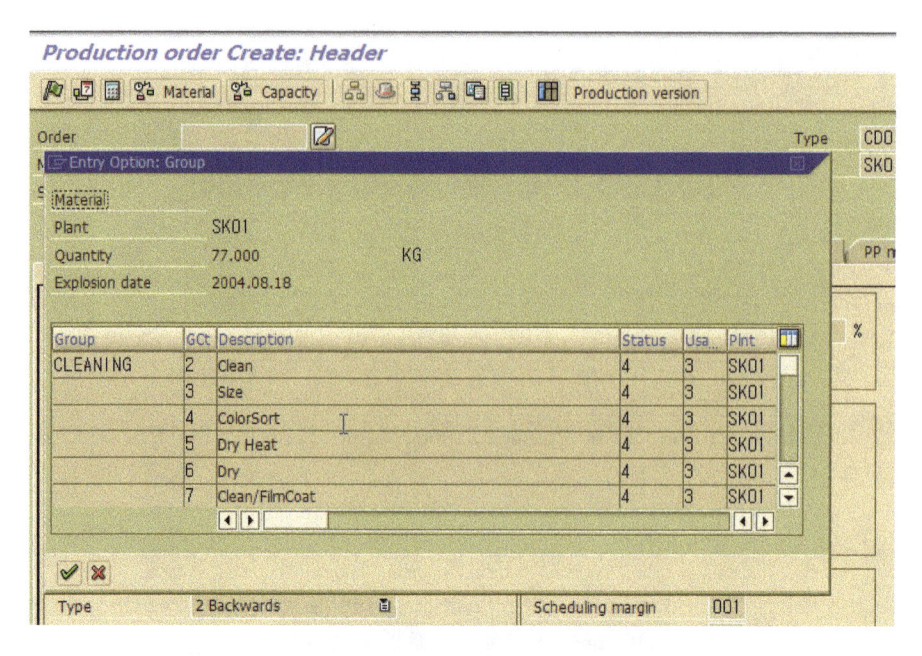

- Choose and double click to apply.

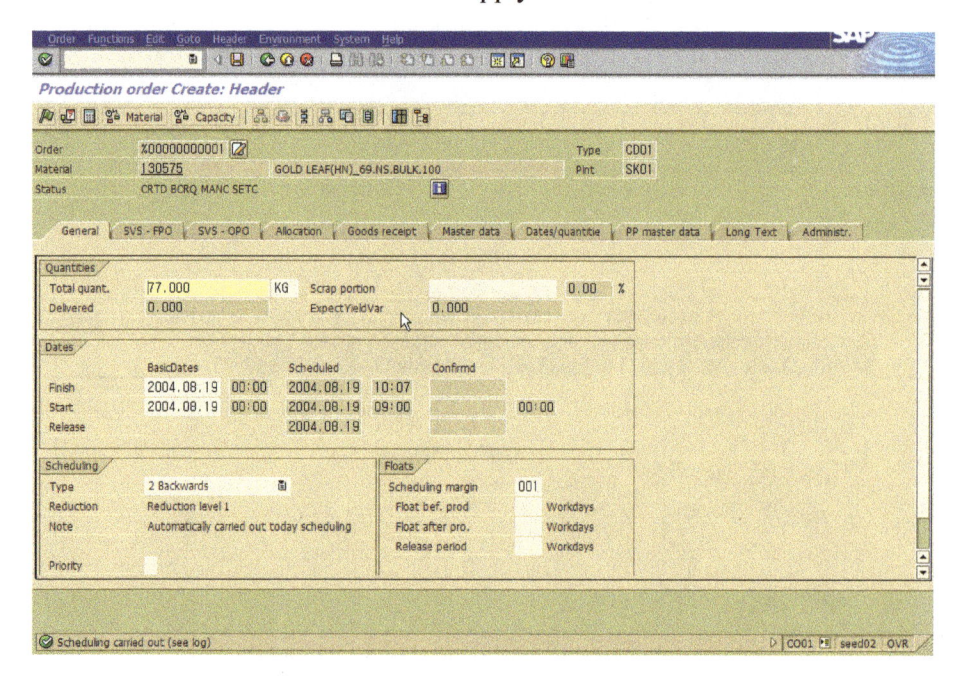

- Confirm Operation overview

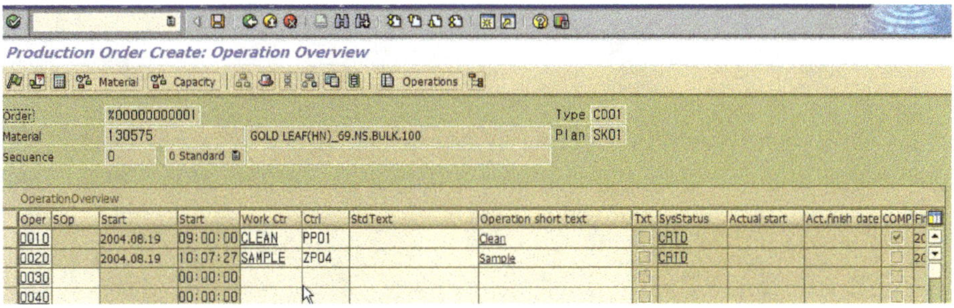
Production Order Create: Operation Overview
Material
Capacity
Operations
Order %00000000001
Type CD01
Material 130575 GOLD LEAF(HN)_69.NS.BULK.100
Plan SK01
Sequence 0 0 Standard
OperationOverview
Oper SOp Start Start Work Ctr Ctrl StdText Operation short text Txt SysStatus Actual start Act.finish date COMP
0010 2004.08.19 09:00:00 CLEAN PP01 Clean CRTD
0020 2004.08.19 10:07:27 SAMPLE ZP04 Sample CRTD
0030 00:00:00
0040 00:00:00

- Check component materials and batches

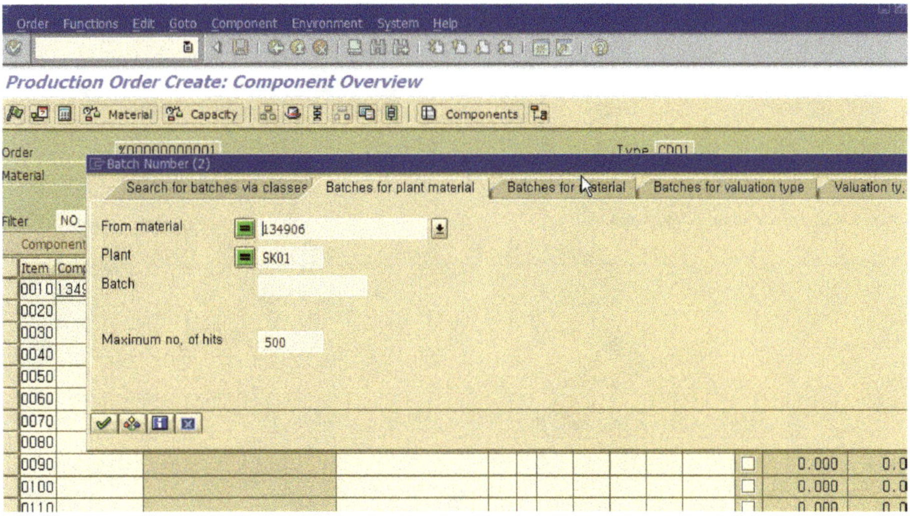
Order Functions Edit Goto Component Environment System Help
Production Order Create: Component Overview
Material
Capacity
Components
Batch Number (2)
Search for batches via classes
Batches for plant material
Batches for valuation type
From material 134906
Plant SK01
Batch
Maximum no. of hits 500

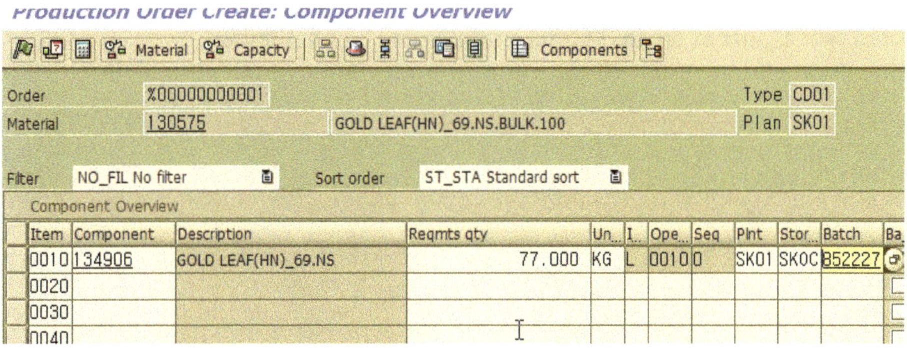
Production Order Create: Component Overview
Material
Capacity
Components
Order %00000000001
Type CD01
Material 130575 GOLD LEAF(HN)_69.NS.BULK.100
Plan SK01
Filter NO_FIL No filter
Sort order ST_STA Standard sort
Component Overview
Item Component Description Reqmts qty Un.. I.. Ope... Seq Plnt Stor... Batch
0010 134906 GOLD LEAF(HN)_69.NS 77.000 KG L 0010 0 SK01 SK0C 852227
0020
0030

- Check and enter

118   Vegetable Seed Processing

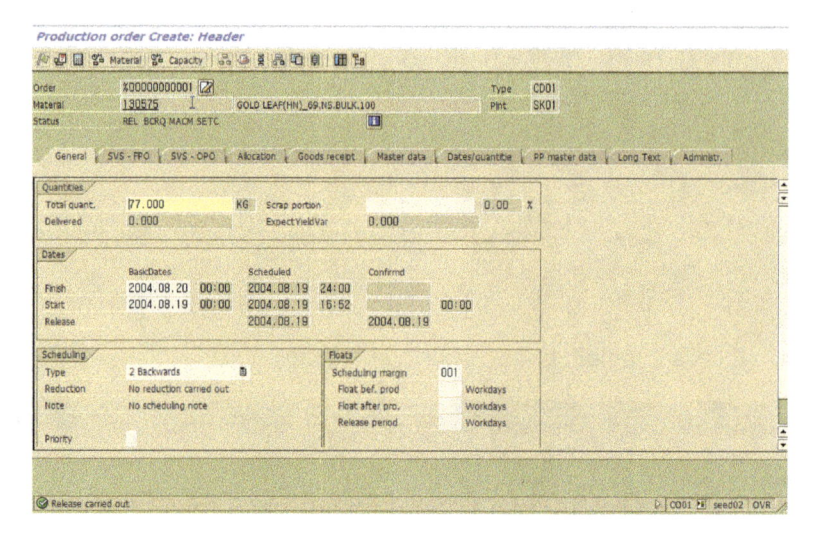

- Release and save

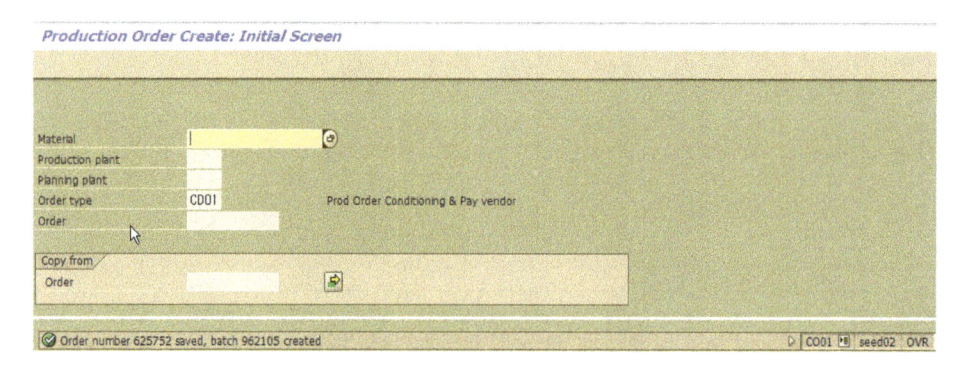

- Confirm created Order number and batch number.

CHAPTER - 11

Guidelines for Quality Vegetable Seed Production

11.1 Planning

- Upon receipt of production program verify that crop sowing dates fall into crop windows identified for the particular area and allot accordingly. Check the quality of lots produced in previous season, if available.
- Assign crop location (Open field, Green house, Insect Proof Net / Cage etc) according to crop needs, site conditions and degree to which these conditions can be manipulated efficiently to meet the crop needs. Always adhere to optimum crop conditions.
- Identify material and personnel needs based on the production program and coordinate timely arrival of both. Material and personnel must meet standards established by the production site.
- Identify material quality like A, B, D (First, Second and Dirty) based on the specs.
- Refer standard procedures or formulate some of them. Do not proceed if proper documentation does not exist.
- Plan training for laborers regarding crossing, flower morphology, anthesis, time of stigma receptivity etc.

11.2 Stock seed handling

- Upon receipt of stock seed or foundation seed for production check details regarding lines to be used as male or female or whether it is an OP.
- Verify whether seed passed appropriate disease testing or assays.
- Verify seed origin-batch no and lot information. (Fig. 77)

Fig. 77 : Batch and lot information

- Check for any particular phytosanitary instructions
- Always use stock seed from specified batch or lot. Never use remnant stock seed for production. If at all stock seed left after sowing return to the foundation seed department.
- Do not hand over stock seed directly to growers or nursery organizer

11.3 Greenhouse/Insect proof net production

- Consult production manual for specific cultural, developmental and sanitation requirements specific to the crop being handled
- Ensure integrity of source seed. Verify that seeding area is free of any foreign seeds before sowing is initiated.
- Never have more than one variety's seed envelop open at a time at a sowing station.
- Return remnant seed to warehouse
- Only one line may be sown at a station at any one time.
- Follow all preventive measures outlined to avoid contaminating the production with possible seed borne diseases.
- Maintain a minimum of 1 meter between varieties of same species.
- Maintain a particular number of plants in the green house and insect proof net. (Fig. 78)

Fig. 78 : Greenhouse seed production

- Transplant orders are to be filled one at a time
- Field inspection stakes must be in place before planting
- Proper land preparation to be done to avoid soil borne diseases and provision for proper drainage.
- Go for drip irrigation system in insect proof net production.
- Plant male and female plants in different rows. Male and female lines should be demarcated with proper tags. (Fig. 79)
- Proper staking should be done for plant to avoid collapsing
- Maintain a specified no of shoots and always prune off the lower leaves & shoots to avoid coming in contact with the soil
- There should be one separate chamber for laborers for sanitation before entering in to the main area in the green house or net.
- Always the seed washing area should be away from the net. (Fig. 80)
- There should be defined spray schedules as per work plan. The dates for various operations should be chalked out before transplanting it self.
- Work plan should consist of details from date of sowing till date of harvest.
- Always display instructions to laborers regarding safety instructions, and disinfectant details before entering in to the net (Fig. 81)

122   Vegetable Seed Processing

Fig. 79 : Proper labeling of lines

Fig. 80 : Insect Proof Net Production in Hot Pepper

Fig. 81 : Safety instructions

11.4 Direct sowing

- Minimum isolation distances to be maintained when direct sowing. For crops that are not at risk of cross pollination, never sow two varieties of a same species without at least a skip (staggering), a physical barrier or a crop of a different species in between. (Fig. 82)

Fig. 82 : Direct and Open field sowing

- Never direct seed into field that contained a crop of the same species the previous cycle.
- Consult production manual for specific cultural, developmental and sanitation requirements specific to the crop being handled.
- Follow all preventive measures outlined to avoid contaminating the production with possible seed borne diseases.
- Field ID stakes labeled with the particular production order no. and line identification must be in place prior to initiating sowing process.
- Sowing information to be documented immediately on field map as soon as sowing completed.
- Follow all the spacing requirements and male-female ratios
- Follow staggering planting according to the varietal as well as line characters
- If mechanized sowing is adopted clean the equipment before and after sowing.

11.5 Transplanting in Net or open field

- Consult production manual for specific cultural, developmental and sanitation requirements specific to the crop being handled.
- Follow all preventive measures outlined to avoid contaminating the production with possible seed borne diseases.
- In situations where crop isolation enters into conflict with the crop rotation, crop isolation will take preference with the strict condition that the crop must transplanted and not direct seeded and prior to transplant the soil must be removed of any volunteer plants.
- Isolation requirements and phyto sanitary conditions must be taken into consideration before assigning a crop to a given area.
- Field inspections must be conducted through out the crop cycle to ensure that the quality of the crop is not at risk at any given moment.
- Inspections must take into consideration neighboring crops as well as the area itself.
- Due to the possible phyto sanitary risk associated with tobacco smoking and chewing should not be allowed in the field and during working hours.

11.6 Roguing (Field inspection)

- Quality assurance personnel and supervisors must inspect all the crops beginning at seedling stage to maturation for possible off types, selfed plants in crossed row, plants showing virus symptom which are not easily controllable must be recorded and rogued out to ensure quality seed devoid of genetic contamination. (Fig. 83)
- If more than 2% off types are detected within production, the stock seed specialist should be consulted.
- Record and documentation of the possible contaminating plants help in future production and PGO testing.
- Also this information helps in compensate the growers.

Fig. 83 : Roguing operation to search for the selfed plants and any offtypes.

Weed Control

- A strict weed prevention and control schedule to be prepared.
- The main field should be always free from noxious weeds of whatsoever type.
- Weeds act as alternative hosts hence should be controlled. Also they compete with seed crop for nutrients depleting the seed quality.

Pollination or Crossing

- Start flower collection for pollen extraction from male parent and crossing at specified time. Since at this period the stigma receptivity is more.
- Train the laborers regarding flower type, time of pollen extraction and crossing time
- Avoid movement of personnel from one crop in flower to another to prevent movement of pollen.

In case of male parent

1. Roguing or removing of off types before start of the crossing is a must.
2. No crossing before Roguing in the male lines.
3. If the pollen quantity is more remove excess male plants.

4. Do not spray flowering inducing chemicals in female lines-this will results in more self plants in female lines.
5. Pollen must be collected in clean container and use pollen-ring for pollination
6. Pollen quality affected by exposing to strong sunlight but a mild early in the morning sunlight induce more an thesis-there by more pollen quqntity.
7. All the lines must be properly separated and identified through out the growing season.
8. During roguing compare the characters of fruits, plants, according to character list provided by the breeder.
9. No more than one male parent may be harvested at any time by any one person or group of person.
10. All containers for collecting pollen should be labeled with production order no. Use color codes in case of CMS parents of Hot pepper
11. During pollen extractions disinfect hands and equipments (like tweezers) with alcohol.
12. When transporting pollen, care must be taken so as to avoid spills or cross contamination.
13. After crossing was over rogue out male plants immediately to avoid contamination with the seed or female parent.

- **In case of female parent**
  1. All the lines must be properly separated and identified through out their growing season. All the lines must be labeled with production order no.
  2. Consult production manual for specific cultural, developmental and sanitation requirements specific to the crop being handled. Proper nutrition to the seed parent is a must. Both macro as well as micro nutrients should be applied as per work plan (Fig. 84).
  3. Follow all preventive measures outlined to avoid contaminating the production with possible seed borne diseases.
  4. In case of emasculation activity should be taken before anthesis is reached. Anthers must be removed completely during emasculation.
  5. Use tweezers for emasculation.

Guidelines for Quality Vegetable Seed Production 127

Fig. 84 : Calcium deficiency in tomato

6. In case of emasculation activity should be taken before anthesis is reached. Anthers must be removed completely during emasculation.
7. Use tweezers for emasculation
8. Remove open flowers in the plants before starting of crossing which otherwise lead to selfing (Fig. 85)

Fig. 85 : Selfed fruits in tomato due to improper emasculation.

9. Always follow specified no days for crossing which varies with each crop
10. After the crossing weed control and proper nutrition helps better quality seed which improves germination.

11.7 Maturation or Harvesting

- Consult production manual for specific cultural, developmental and sanitation requirements specific to the crop being handled. Proper nutrition to the seed parent is a must. Both macro as well as micro nutrients should be applied as per work plan.
- Follow all preventive measures outlined to avoid contaminating the production with possible seed borne diseases.
- At least 10 days before harvest remove any non hybridized fruits (Selfed)
- Check for harvest colour as in case of Hot pepper. There is correlation between quality and harvesting at specific colour percentage.(e.g at 70%,80%,90%,100% etc.) (Fig. 86)

Fig. 86 : Harvest at 90 % colour development

- Check for tags as in case of bitter gourd-Harvest only tagged fruits to ensure purity.
- Only healthy fruits that are free of any disease should be harvested. Before start of the harvesting discard or remove diseased, bird affected, rotten, insect infested fruits. The seed from these seed if mixed with good seed decrease germination. (Fig. 87 & 88)

Guidelines for Quality Vegetable Seed Production   129

Fig. 87 : Rotten and bird affected fruits

Fig. 88 : Diseased fruits

- All harvesting, seed extraction, washing and drying equipment must be checked to verify that it is free of any foreign seeds prior to initiating that particular process.
- Always harvest specific time as per crop profile sheet.

11.8 Seed receipt at processing plant

- No seed will be accepted in the processing plant without harvest tags one inside and one outside of the bags (Fig. 89).

Fig. 89 : Harvest Tags

- Each lot should be written with all the details on the bags (Fig. 90)

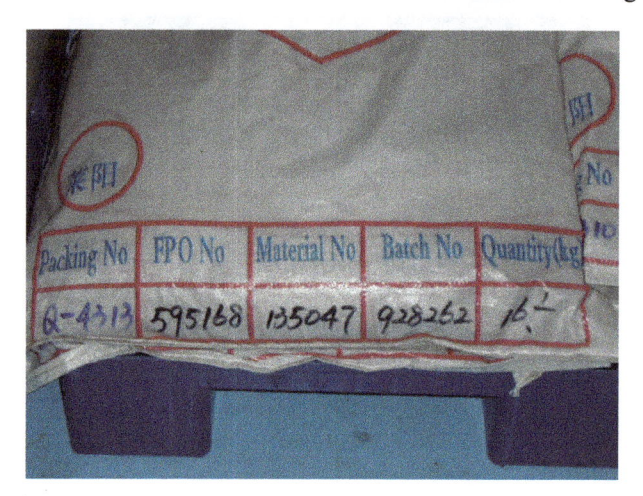

Fig. 90 : Details on seed bag at processing plant

- While taking seed sample for tests verify seed probe and work area is free of foreign seeds before initiating sampling process

Fermentation

- Fermentation is a method to remove seed from the pulp and also to prevent seed borne diseases. This method followed in case of Tomato, Bittergourd, Gourds, Eggplant and Watermelon. Care should be taken not to allow water to come in to contact with fermenting product which reduces germination.

Guidelines for Quality Vegetable Seed Production   131

Seed Washing and Treatment

- Always use clean water for seed cleaning and washing
- The washing process must be done as quickly as possible to avoid activating the germination process within the seed. The seed must never be submerged in water for more than 1 hour.
- In case of Tomato, Hot pepper seed wet seed treatment with H2SO4 & Calcium Hypochlorite is recommended to prevent seed borne pathogens (Fig. 91).

Fig. 91 : Wet Seed Treatment

Seed Drying

- Identification tags must accompany the seed through out the seed drying process.
- At the production site shade drying above the soil is preferred (Fig. 92). Avoid direct sun drying.

Fig. 92 : Temporary shade drying arrangement in the field

- In case of machine drying in the processing plant equipment must be checked to confirm that it is free of any foreign seeds before placing for drying. Always follow the instructions setting proper temperature and duration. (Fig. 93)

Fig. 93 : Drying using equipment

Seed Cleaning and Blending

- All seed taken for cleaning must be labeled and each equipments like screens should be cleaned before start of cleaning. (Fig. 94)

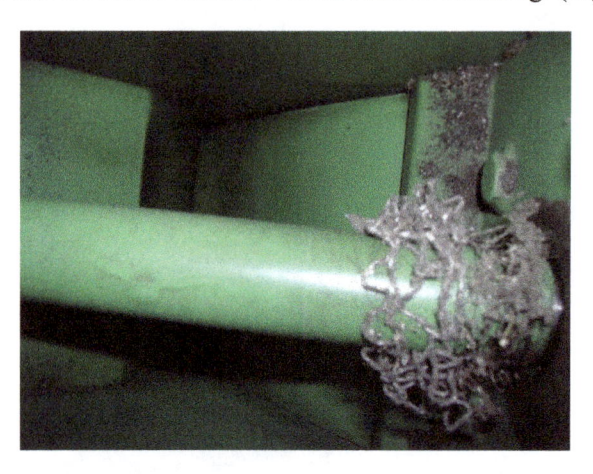

Fig. 94 : Cleaning of cleaning equipments

- Care should be taken while taking seed lots for blending, Based on the physical appearances also we can differentiate the seed lots to avoid contamination. (Fig. 95)

Guidelines for Quality Vegetable Seed Production   133

Fig. 95 : Physical appearance for blending seed lots

- Blend only specified lots as per reports of QA and Production departments. Never blend a high standard lot with low standard one. There should be only 10 or less points difference.

Seed Packaging

- Before initiating the seed packing process ensure that both the work areas and the containers are free of foreign seed. Follow proper container size, amount.(Fig. 96)

Fig. 96 : Clean surroundings while packing manually & Stacking in seed storage

Seed Storage

- Seed warehouse must comply with safe storage like Rh, Temperature, seed moisture etc
- Unidentified seed lot must not be stored in the warehouse
- Observe maximum stacking heights in order to avoid seed spillage

Table 10 : Production Areas for Vegetable Seed Production

Sl no	Area (in Karnataka)	Vegetable Seed	Remarks
1.	Haveri District- comprising Ranebennur, Haveri, Byadagi, Hirekerur,	Tomato, Hot pepper, Eggplant (Brinjal), Capsicum, Cucumber, Gourds (All Type)	Hot pepper and Capsicum under insect proof net
2.	Koppal	Watermelon, French Beans	
3.	Shira	Muskmelon, Squash	
4.	Shiggaon and Savanur	OP-Okra	
5.	Chitradurga	OP-Onion	Both Bulb and Seed
6.	Mundaragi	Bittergourd	
7.	Arasikere	Hot Pepper	
8.	Davanagere	Gourds	
9.	Shimoga	Hybrid Okra, Coriander	

Table 11 : Pollination or Crossing Timings

Sl no	Crop	Time of Crossing	Pollen Collection
1.	Tomato, Hot pepper, Eggplant, Okra, Bitter gourd, Capsicum	Morning-8 am to 2 pm	Evening 6 pm to 8 pm
2.	Cucumber, Watermelon	Morning-8 am to 12 pm	Evening or in morning
3.	Squash, Musk melon	Morning-4 am to 7 am	Evening or in morning

Chapter - 12

Glossary

A

After-ripening. The physiological maturation processes which occur in e.g. seeds and fruits after harvest or abscission. After-ripening is often necessary for immature seeds to become germinable. Also used for the seed handling process itself.

Application Rate: The application rate of Seed Treatments, Seed Treatment Products and/or Seed Treatment Components. For liquid Seed Treatment Products, rates are typically expressed in ml or in mg ai/kernel on the label.

Aspirator or blower: A processing equipment in which when a seed mixture is introduced in to a upward moving air stream, light particles with a low terminal velocity are carried along with the flow where as heavy seed falls through.

B

Batch: A defined quantity of seed material, packaging material or product processed in one process or series of processes so that it could be expected to be homogeneous.

Bio-priming : is a process of biological seed treatment that refers combination of seed hydration (physiological aspect of disease control) and inoculation (biological aspect of disease control) of seed with beneficial organism to protect seed. It is an ecological approach using selected fungal antagonists against the soil and seed-borne pathogens. Biological seed treatments may provide an alternative to chemical control.

Batch number (or lot number): A distinctive combination of numbers and/or letters which specifically identifies a batch.

Bulk Seed: Any seed lot or batch which has completed all seed processing stages up to, but not including, final packaging.

Blending: It is simply combining two or more seed lots of same variety or hybrid of specified quality. The blending may involve lots which may have different quality in terms of germination but the blended lot meet the minimum standards.

Bucket Elevator: Conveying equipment used in processing to more seed to overhead storage bins, designed to cause minimal mechanical damage to seed.

C

Calibration: The adjustment of seed treatment equipment to apply the target rate of slurry or Seed Treatment Product and the verification thereof.

Chaff: Thin dry bracts or scales, especially the dry bracts enclosing mature grains of cereal grasses, primarily removed during threshing but which may be present in low levels in commercial seed. More generally, seed debris.

Class (of seed and seed crop) – Refers to the generations of pedigreed seed and seed crops, such as Breeder, Foundation, Certified and truthfully labelled seed which have met the standards prescribed by recognized seed and seed crop certification agencies.

Cleaning, pre-(seed): Separation of seed from other species and non-seed fragments such as fruit fragments, leaves or stems. Cleaning may be undertaken by sifting, blowing, winnowing, flotation etc.

Colour Sorter: A processing equipment which removes defective individual seeds one at a time based on differences in reflected colour or brightness under one or two selected visible or infrared light wavelengths.

Conditioning of seed – A term used to describe the cleaning of seed, usually to improve mechanical purity.

Cryopreservation or cryo-storage. Maintaining tissues or seeds for the purpose of long time storage at ultralow temperature, typically between -150°C and -190°C. In normal cryopreservation the sample is pre-treated with a cryo-protective substance, followed by slow, controlled freezing. This method is often employed for recalcitrant seeds.

D

Dis-infection, seed: It refers to eradication of fungal spores present within the seed coat or more deep seated tissues. For effective control the fungicide must penetrate into the seed to kill the fungus.

Dis-infestations, seed: It refers to the destruction of surface borne organisms that contaminated the seed surface but not infected the seed. Chemical dips, soaks, sanitation, fungicides applied as dust, slurry or liquids have been found successful.

Desiccant: Chemical compound that has a high moisture absorption affinity and can be used for desiccation or maintaining a low humidity when stored together with e.g. seeds. Common desiccants are SiO, CaO.

Dry seeds: This type of crops like sponge gourd, ridge gourd, melons, onion, radish, carrot, spinach or most of the exotic vegetables. Here the seed is not attached to flesh at the time of harvesting. The flesh holding the seed will be either completely dried or seed bearing on the dry umbels like in onion

E

Enhancement, seed: Seed treatment to improve lustre, quality and performance like polymer coating or priming. Also known as functional seed treatment.

Equilibrium moisture content (EMC): Moisture content of seeds in equilibrium with atmospheric humidity at a given temperature. The EMC is influenced by hygroscopic character of the seed storage material viz. low for oily seeds and high for seeds rich in protein and carbohydrates

Equilibrium Relative Humidity (ERH): The percentage relative humidity at which a given EMC is expressed at a specified temperature.

Encrusting, seed: encrusting applies to less material so the original seed shape is still visible.

F

Fluming or flotation: It is a method of removal of light weight and immature seed by floatation in water as part of seed extraction activity.

Film coating: Seed treatment with materials which regulate seed water availability and gas exchange, there by controlling timing of germ and emergence.

G

Genetic purity – Trueness to type or variety.

Grading: Separation of seed crop produce in to different categories according to size, weight, colour and quality.

H

Handling seed: Handling includes the movement or products and treated seed, including but not limited to loading, unloading, weighing, bagging, sewing, stacking, and planter loading and operation.

Hardening, seed: It is a process of seedling grown under green house or shade net before transferring to harsher field conditions by reducing watering, shelter and fertilizer & so on.

Hazardous components: Components which present health, safety or environmental hazards.

Hermitic storage: Storage inside air-tight containers. This is very essential for maintaining low moisture content required for successful long term storage of dried orthodox seeds.

Hygroscopic: Sensitive to moisture, absorbing or losing water. Pores on seed coat open and close according to humidity because they contain hygroscopic tissue.

I

Imbibition: The process of initial water uptake by seeds prior to germination. Imbibition is an entirely physical process and also non-viable seeds imbibe.

L

Labels – issued to identify the crop, variety, standards and class of a seed lot. Sometimes called "tags."

Liquid Density Seperation: Method for cleaning seeds from particles with higher or lower specific density by submerging in water or other liquid. Used e.g. for separation of empty and filled seed and mechanically damaged seed

Longevity, seed: The period of time seed will maintain viability in storage under a given set of storage conditions. Often used equivalent to storability

M

Material Safety Data Sheets (MSDS): The MSDS is a detailed informational document prepared by the manufacturer or importer of a hazardous chemical. It describes the physical and chemical properties of the product, and is a tool for communicating safe handling and environmental protection requirements for chemical products. For pesticide products such as Seed Treatment Products, the information on the label takes precedence over the information on the MSDS.

Moisture meter: Instrument for quick measurement of moisture content of seed without drying. The instrument measures the electrical properties of the seed tissue, which is correlated with the content of water (moisture content). Moisture meters must be calibrated for each species with standard method of moisture content measurement i.e. oven-drying.

N

Noxious weed – A weed or plant that is considered undesirable and so categorized by Seed Act or Rules or Indian Minimum Seed Standards

O

Orthodox: Term used to describe seeds which can be dried down to a low moisture content of around 5% and successfully stored at low or sub-freezing temperatures for long periods. There is practically no metabolism in dry, cooled orthodox seeds, but they may deteriorate by general ageing and thus ultimately lose their viability

Other crop seed – One of the four components of a seed purity test and usually refers to the number of seeds of other crop kinds in the seed sample being tested.

P

Packaging: The container holding the untreated or treated seed or holding Seed-normally a aluminium foil or a tin

Personal Protective Equipment (PPE): Equipment that is worn by employees to mitigate hazards of a process. For seed treating operations, PPE is typically means to reduce exposure of operators to seed treatments and treated seed dust. Such PPE includes but is not limited to long-sleeved shirts; long pants; shoes; socks; goggles; chemical resistant gloves; and respirators.

Pelleting: Procedure by which individual seeds are provided with an envelope of adhesive material containing e.g. nutrients, microsymbiont inoculant and/or pesticides. In addition to providing these beneficial compounds, pelleting facilitates mechanical sowing because of the more uniform seed size

Physical purity – Refers to the degree of freedom of a seed lot from seeds of other crop kinds, weed seeds and inert matter.

Physiological maturity: General term for the stage in the life cycle of a seed when development is complete and the necessary biochemical components for all physiological processes are active or ready to be activated.

Priming: Pre-treatment method to promote rapid and uniform germination. The seeds are soaked in a liquid solution (e.g. polyethylene glycol (PEG), sugar or salt) of sufficiently low water potential to regulate moisture content at a level where the germination process initiates but radicle protrusion is prevented.

Processing, seed: Seed handling methods from collection to storage, usually a collective term applied to extraction, cleaning and drying.

Protection, seed: To protect the seed and young seedling from organisms in the soil which might otherwise cause delay of the seed before germination.

Purity: Proportion of clean, intact seed (according to pure seed definition) of the designated species in a seed lot, usually expressed as a percentage by weight.

R

Recalcitrant: Term used to describe seeds that cannot survive drying below relatively high moisture content (30-40%) and, for tropical species, do not tolerate low temperature. The seeds rapidly lose their viability and cannot be successfully stored for long periods

Relative humidity (RH): The actual amount of water vapour in the atmosphere as a percentage of that contained in an atmosphere saturated with water at the same temperature.

S

Sampling – The method by which a representative sample is taken from a seed lot to be used for analysis.

Sample: In the context of seed testing, a small representative quantity drawn from a seed lot. The different types of samples in seed testing are primary sample, composite sample, submitted sample and working sample.

Seed Flow: The uniformity and freedom of flow of seed through a system, generally through a seed conditioning or treating plant; or through a planter. Poor seed flow may be slow or inconsistent seed flow, or plugging of auguer or conveyors or other handling equipment. Seed treatments may positively or negatively impact seed flow.

Seed lot: A specified quantity of seed of the same species, provenance, date of collection and handling history, and which is identified by a single number in the seed documentation system.

Seed lot system: A seed lot system is a system according to which successive batches of a product are derived from the same master seed lot at a given passage level. For routine production, a working seed lot is prepared from the master seed lot. The final product is derived from the working seed lot and has not undergone more passages from the master seed lot than the vaccine shown in clinical studies to be satisfactory with respect to safety and efficacy. The origin and the passage history of the master seed lot and the working seed lot are recorded.

Seed Moisture: The free water content of seeds, typically measured as a percentage.

Seed Production : The process of growing crops to be sold as seed, instead of growing crops to sell as food or feed. Because each seed will become a crop plant, seed production requires high standards for quality and germination

Seed Treater: Equipment designed to apply seed treatments to seed. Such equipment should be designed so it can be calibrated to accurately and uniformly apply the product to seed. Numerous types of seed treaters exist.

Seed Treatment: Seed treatment is the application of biological organisms and chemical ingredients to seed to suppress, control, or repel plant pathogens, insects, or other pests that attack seeds, seedlings or plants. Seed applied technologies such as inoculants, herbicide safeners, micronutrients, plant growth regulators, seed coatings, colorants, etc. may also be applied to the seed. Treated seed is intended for planting only and not for food or feed uses.

Seed Treatment Polymers: Products added to seed treatments whose primary function is to reduce dust of treated seed and to improve retention of seed treatment active ingredients on the seed

Screens - Screens with different-sized openings are used to separate seeds from chaff and distinguish seeds. The screen number denotes the number of openings that will cover a one inch line. A screen is selected with openings just large enough to let seeds drop through without the chaff or as in the case of larger seeds, a screen selected to allow the chaff to drop through without the seeds.

Slurry: The combined treating composition for application to seed. It may be as simple as a single ready to use product, or a combination of several Seed Treatment Components and water.

Sorter: Pneumatic sorting tables are used to separate seeds by weight, and a grain cleaning column is used for cleaning seeds from other substances with different aerodynamic properties. Gravity seed cleaners and electromagnetic seed cleaning machines sort the seeds by shape and surface texture.

T

Threshing: Disintegration and extraction of seeds from dry fruits by mechanical impact to the fruit, e.g. flailing, beating, trampling, stamping, or by threshing machines.

Treated seed: Seed that has been treated with a "Seed Treatment Product" like fungicide, pesticide, polymer or any seed quality enhancing material.

U

Upgrading: Improving the average quality or performance by removing inferior individuals. In connection with seeds, the increase of viability and vigour of a seed lot by removal of small, immature, empty and otherwise inferior seeds from a seed lot. In connection with seed sources, the culling or roguing of inferior phenotypes to improve the genetic quality.

Ultra-dry storage: Seed storage in conditions where low temperature storages are not available. Here seed is dried to <5% moisture and equilibrium at 10-12 RH% and temperature at 20°C

V

Vigour – The vitality or strength of germination especially under unfavourable conditions.

W

Wet seeds- This type of seed like tomato, peppers, egg plant, capsicum, watermelon, cucumber, bitter gourd, bottle gourd or pumpkin, where the seed is attached to flesh of the fruit and seed separation will be based on fermentation principle

Winnowing - An ancient technique used to clean seeds moving air from a fan or breeze is used to separate heavier seeds from lighter chaff.